D1486769

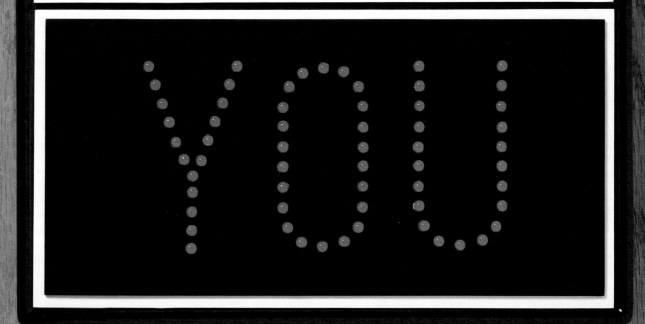

TIME
The Science of You

TIME

MANAGING EDITOR Richard Stengel
DESIGN DIRECTOR D.W. Pine
DIRECTOR OF PHOTOGRAPHY Kira Pollack

The Science of You
The Factors That Shape Your Personality

EDITORS Stephen Koepp, Neil Fine
DESIGNER Sharon Okamoto
PHOTO EDITOR Dot McMahon
WRITERS David Bjerklie, Laura Blue, Michael Q. Bullerdick, John Cloud,
Henry Kellerman, Jeffrey Kluger, Michael D. Lemonick, Belinda Luscombe,
Alice Park, Joel Stein, Maia Szalavitz, Sherry Turkle, Bryan Walsh
COPY EDITOR David Olivenbaum
REPORTERS Ellen Shapiro, Jenisha Watts
EDITORIAL PRODUCTION Lionel P. Vargas

TIME HOME ENTERTAINMENT
PUBLISHER Jim Childs
VICE PRESIDENT, BRAND AND DIGITAL STRATEGY Steven Sandonato
EXECUTIVE DIRECTOR, MARKETING SERVICES Carol Pittard
EXECUTIVE DIRECTOR, RETAIL AND SPECIAL SALES Tom Mifsud
EXECUTIVE PUBLISHING DIRECTOR Joy Butts
DIRECTOR, BOOKAZINE DEVELOPMENT AND MARKETING Laura Adam
FINANCE DIRECTOR Glenn Buonocore
ASSOCIATE PUBLISHING DIRECTOR Megan Pearlman
ASSISTANT GENERAL COUNSEL Helen Wan
ASSISTANT DIRECTOR, SPECIAL SALES Ilene Schreider
DESIGN AND PREPRESS MANAGER Anne-Michelle Gallero
BRAND MANAGER Michela Wilde
ASSOCIATE PRODUCTION MANAGER Kimberly Marshall
ASSOCIATE BRAND MANAGER Isata Yansaneh
ASSOCIATE PREPRESS MANAGER Alex Voznesenskiy

EDITORIAL DIRECTOR Stephen Koepp

SPECIAL THANKS TO:
Katherine Barnet, Jeremy Biloon, Susan Chodakiewicz, Rose Cirrincione,
Jacqueline Fitzgerald, Christine Font, Jenna Goldberg, Hillary Hirsch, David Kahn,
Amy Mangus, Amy Migliaccio, Nina Mistry, Dave Rozzelle, Ricardo Santiago,
Adriana Tierno, Vanessa Wu, TIME Imaging

ISBN 10: 1-61893-056-7
ISBN 13: 978-1-61893-056-9
Library of Congress Control Number: 2012955296

We welcome your comments and suggestions about TIME Books. Please write to us at:
TIME Books, Attention: Book Editors, P.O. Box 11016, Des Moines, IA 50336-1016.
If you would like to order any of our hardcover Collector's Edition books, please call us at 1-800-327-6388,
Monday through Friday, 7 a.m. to 8 p.m., or Saturday, 7 a.m. to 6 p.m., Central Time.

Some of the articles in this book previously appeared in TIME or on TIME.com.

Contents

90
Dynamics
Can we change our habits? Transforming our behavior means fighting the brain's natural orientation. That's certainly possible, but it isn't painless.

52
Types
Are you a dreamer, a doer, a protector? Or just the type who loves personality tests? If you put yourself in that last category, you're not alone.

110
Essay
One fearless man's mission to prove that humor can be taught leads him to ask the probing question: is being funny 28% genetic?

68
Disorders
For many of those with conditions like narcissism or borderline personality, the most difficult step can be acknowledging their problem.

What Shapes Us

The greatest mysteries of all may be the ones that lie within.

By Jeffrey Kluger

THE PEOPLE WHO STUDY HUMAN BEHAVIOR OUGHT TO BE the laughingstocks of the science world. Physicists, after all, at least get answers. It may take half a century, billions of dollars, and an accelerator the size of a small city, but ultimately their theories get either a thumbs-up or a thumbs-down. Does the Higgs boson exist, someone asks in 1964. Yes, it does, comes the answer in 2012.

The same kind of certainty is possible for the biologist and botanist and astronomer. There is always a protein or plant or planet that is either out there to be found or not, each of them one more chance for intrepid scientists to check off a box in their research and move onto the next. But that is not the case for the hardworking people who study the human mind. They labor in a place that is science's hall of mirrors, where shards of answers are revealed, others are held back, and others are sheer illusion. Yes, there are hormones linked to depression, neurotransmitters for bliss, clumps of neural tissue that light up or shut down when we feel rage or anxiety or grief. But there's more too, even if we don't yet know what it is.

We know which areas of the brain go quiet and which start to churn in the dark of night, for example, explaining why we can be so wonderfully creative in our sleep. Still, only the luckiest few wake one morning with the inspiration for a symphony in their heads. The rest of us just dream about flying fish or duck-billed horses, and that's the end of it. Our personalities may be three parts genes, two parts chemistry, and one part learning, but they're many parts secret sauce too, and try as we might, we've never been able to figure out the complete recipe.

That, however, is the secret joy of the mind scientists' work—the journey away from crisp questions and straight answers, and into the strange, near-hallucinatory world within. In some ways, it's a journey we are all making right along with them.

You inhabit yourself from the moment you're conceived, and yet you never quite know how to navigate your internal terrain. You're shaped by nature and nurture, by culture and family; we all are. The emotional ecosystem that results makes us capable of transcendent love, of wild creativity, of profound wisdom, of deep empathy—and also of depression, obsessions, addictions, rage, and paralyzing fear. For the life of us we can't figure out where all of that comes from.

The answers to those puzzles are as essential as they are elusive—the fundamental particles not of the universe or of mere matter but of us. We will never heal psychological pain if we don't know what causes it, never control human savagery if we don't know why it exists in the first place. And we will never satisfy one of our most primal needs—the need to learn—if we don't allow the mind to turn its attention to the most basic thing of all: itself. So don't laugh at the scientists who ask these questions. They have a pretty enviable gig, going places no one has gone to discover things the rest of us could never have imagined.

Nature

We begin to reveal our psychological selves before we can walk—some of us, for example, seem predisposed to be happy or caring. But that doesn't mean we owe who we are solely to a few strands of DNA.

{
PERSONALITY IS IN THE GENES
FOSTERING EMPATHY
THE BIOLOGY OF JOY
}

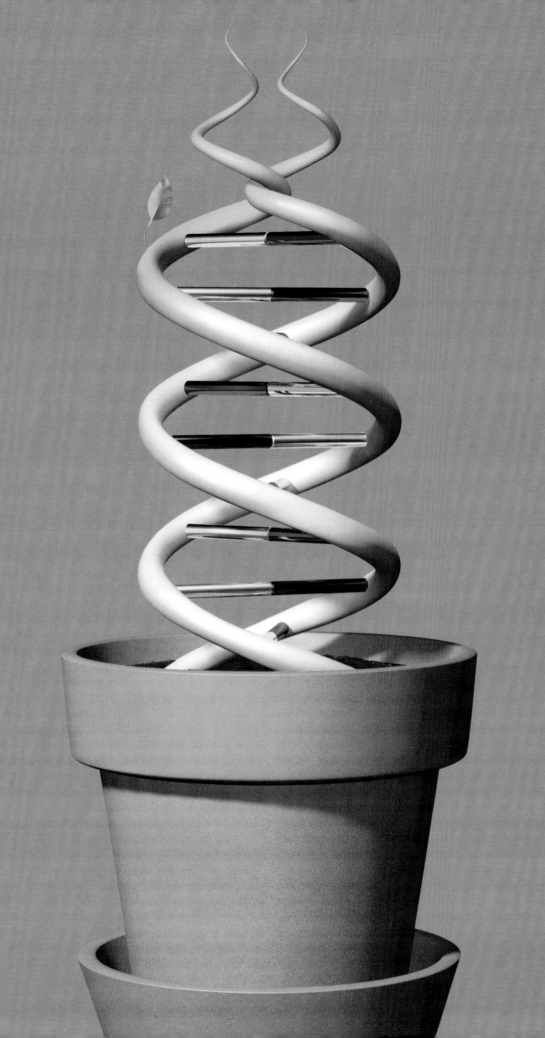

Born to Be Wild

*... Or stubborn, or loving. Yes, genetics does affect
who we are. To a point, anyway.*

By Alice Park

NOT LONG AFTER A NEWBORN joins a family the comparisons start to fly. "He's so happy—just like his mother." "Look at how she loves all the attention; she's just like her father." Our physical features define what we are—tall or short, blonde or brunette—but our personalities characterize who we are: how trusting, how impulsive, how kind or affectionate, how excitable or conscientious or even how likely to feel blue. And they can be so obvious, so early, that everyone already has a fairly good idea of who we are—and who we are like—even before we can walk or talk.

It's what heredity is all about. And there's no question that the genes our parents bestow upon us play a big part in deciding which traits we end up with. The blueprint for every bit of us—from hair color to the shape of our toes to, yes, our personalities—was drawn up pretty much as soon as the baby-making began.

And yet when you consider how essential our psychological features are to how we define ourselves, it seems that the science is not nearly where it should be in understanding the specific mechanisms that create them. One thing that is becoming increasingly plain, though, is that the process, unlike that which governs many of our physical traits, goes far beyond merely matching up the genes in column A to the personality traits in column B.

From the first days out of the womb and on into the toddler and tween years, children are sponges, observing and mimicking the behavior of their elders. How quick we are to anger, for example, may very well have as much to do with how we once saw our parents respond to their own frustrations as it does with whatever specific lengths of DNA code for hot tempers. It's no stretch to see how watching Mom or Dad react with R-rated epithets to breaking a favorite mug could plant that kind of explosive response in their children, while a more measured acceptance—"Whatever, it's just a cup"—could teach members of the next generation to moderate their own anger.

Clearly, the human personality is a well-woven tapestry of DNA-based threads and environmental ones, and teasing them apart would be no easy task under the best of circumstances. And until recently those circumstances were far from ideal, since the research was hampered by science's rather rudimentary grasp of how the gene part of the tapestry worked.

Studies of twins are still the gold standard for sussing out the biological component of personality, and for the longest time it was the best way to glean clues about parents' DNA legacy. The studies consistently showed that identical siblings tend to share personality traits, and as

they also share the same DNA, it made sense to assume that their personalities were genetically rooted. In the most recent study, researchers at the University of Edinburgh in Scotland asked more than 800 sets of identical and fraternal twins about how they perceived themselves and others, then used their answers to slot them into one of the five primary personality categories recognized by the discipline: agreeableness, conscientiousness, neuroticism, extroversion, and openness. Identical twins were twice as likely to share personality traits as fraternal twins, whose genetic similarity is no different from that of nontwin siblings. Adoption studies comparing personalities of siblings who were raised apart backed up these findings; after accounting for the effects of parenting, environmental, and life experiences, researchers determined that anywhere from 35% to 50% of contributions to personality traits could be traced to DNA.

"It's pretty broadly accepted in the field that slightly less than half of the overall variability in temperament is due to genes, while the rest is due to a range of environmental factors," says Kenneth Kendler, director of the Virginia Institute for Psychiatric and Behavioral Genetics at Virginia Commonwealth University. "That's not too controversial at this point."

The research is starting to move beyond these traditional twin studies, to take advantage of genome analyses that can determine exactly where along the complete swath of humankind's 25,000 genes any two people diverge. These genome-wide association studies, as they are known, are fast becoming the best way to identify DNA hotspots—where people suffering from a certain disease, for example, differ from those who don't—and they have already proved invaluable in illuminating the genetic culprits behind everything from Alzheimer's disease to Crohn's disease to psychoses like schizophrenia. Unfortunately, the technique hasn't been as useful in cleanly isolating the roots of personality disorders, for a number of reasons that scientists are just beginning to understand.

Early on, the wisdom went that traits either were solely dictated and regulated by DNA or were the product of external influences alone. Consider the case of vasopressin 1a (AVPR1a), which headlines have dubbed contradictorily as the "ruthlessness gene" or the "fidelity gene." Vasopressin, a hormone released by the pituitary gland, controls the contraction of tiny blood vessels and other muscles. In voles, which are members of the rodent family, the hormone is linked to the bonding between parents and their offspring as well as the attraction between mates. But while meadow voles act like most mammals, mating as many times with as many partners as possible, the bet-

The personality is a tapestry of DNA-based thread and environmental ones. Teasing them apart is no easy task.

{The Optimist}

FROM LEFT: MICHAEL AUSTIN/THE ISPOT; MELINDA BECK/THE ISPOT

ter to perpetuate their genetic legacy, their closely related cousins the prairie voles are unique among mammals for being monogamous. And it just so happens that meadow voles have fewer AVPR1a receptors than prairie voles do. Because vasopressin release is also connected to the brain's reward system, researchers theorized that the brains of prairie voles may be more saturated with good feelings after they mate, and associating that satisfaction with a particular partner prompts them to forge a closer and longer-term bond.

Those findings led directly to more intriguing research in people, which showed that certain versions, or variants, of the AVPR1a gene were also associated with more altruistic behavior. In one study in which participants were given the choice to either keep a $14 reward or pass on portions of it to strangers, those who kept all of the money were more likely to share one version of the gene while those who gave away the entire amount harbored another. Those results suggest that vasopressin 1a, and of course the gene that regulates it, has some link to the extroversion personality trait, such that higher levels of the hormone reinforce the warm feeling of doing good while lower levels make people more selfish.

Similarly, researchers have uncovered a version of a gene coding for an enzyme that breaks down brain chemicals and appears in animals that display more aggressive or violent behavior. And brain-imaging studies with human subjects have shown that those with a similar gene variant were more likely to react strongly, or show more activity on functional brain scans, to emotionally disturbing images than those without that version. That suggests the gene might be involved either in regulating emotional stability or, conversely, in contributing to the neuroticism personality trait. With both of these gene variants, though, the observed connection didn't hold in all cases. And it's this finding that, in its way, is the most significant of all, because it threw into question the very possibility of linking genes to personality.

The most comprehensive look at the genetic drivers of the Big Five personality traits further put the lie to the gene-centric theory. Scientists scanned the genomes of 20,000 people for 2.5 million variants in their DNA, hoping to match specific genetic sequences with one or another of the traits. Only two variants showed even faint links: one to openness, the other to conscientiousness. And who knows if even these weak associations will withstand deeper scrutiny. "It was a start, but not a home run," says James Potash, head of psychiatry at the University of Iowa. "Sort of like a single."

More recent attempts to pin down just how genes are

{The Worrier}

turned on and off might shed some light on such confusing results, because they have led to this game-changing conclusion: the nature-vs.-nurture debate isn't really a debate at all. These days, most researchers have come to believe that nurture actually influences nature in subtle but important ways. "We are not just impacted by our biological programming; our biological programming reacts to the environment too," explains Thomas Lehner, director of the office of genomics-research coordination at the National Institute of Mental Health.

What is important, in other words, is not simply whether a specific gene is present or not, but whether that gene has been set off by some external trigger. It now seems likely that certain versions of genes do not become active until some influence—a stressful experience or trauma, for example—creates the necessary molecular spark that revs the gene into action.

In 2003, Avshalom Caspi, a psychologist with appointments at King's College London and Duke University, created a stir when he and his colleagues identified a genetic variant that was associated with violent behavior, but only among people who had been mistreated as children. The variant, which exists in two forms—one

{The Introvert}

{The Angry One}

inherited from mothers, the other from fathers—codes for a protein that regulates serotonin, the brain chemical responsible for controlling mood and feelings of satisfaction. Previous research had found that people with one or two short versions of the gene tended to score higher on tests measuring neuroticism, specifically accounting for up to 4% of the increased anxiety these people expressed and making them more vulnerable to depression than those with longer versions of the gene.

But when Caspi then tracked 847 people from age 3 for more than two decades, he found an even stronger relationship. Among those who reported four or more stressful life experiences—such as the loss of a loved one—twice as many people with the two shortened versions of the gene recorded symptoms of depression compared with those who had two longer forms of the gene. And among those subjects with a history of child abuse, up to two-thirds of those with two shortened forms of the genetic variant who were abused as children reported feelings of depression, while none of those with two long forms of the gene did.

That kind of two-tiered breakdown represents the latest approach in determining how traits and tendencies of all sorts are produced. It is the centerpiece of a new field known as epigenetics. "People for a long time talked about nature versus nurture as if they were two separate things," explains Potash. "Epigenetics provides the link between the two. It's the way in which nurture influences and ultimately changes nature." So even if children are born with a genetic propensity toward anxiety or moodiness, they won't actually present with either until they've been exposed to a fear-inducing or traumatizing environment, such as being raised in an orphanage or growing up in a war-torn country. If all those earlier studies failed to show a strictly genetic basis of personality, if they were unable to show a direct connection between genetic variants and personality traits, it could very well be because they weren't taking into account this more subtle effect of genes.

Researchers are just beginning to apply this newly embraced epigenetic understanding to personality disorders, in the hope of generating a more sophisticated

FROM LEFT: BRAD YEO COLLECTION/THE ISPOT; KEN ORVIDAS/THE ISPOT model of how personalities of all stripes emerge. Focusing on pathological states provides a window into what goes awry in the intricate and fragile interplay between biology and surroundings, and comparing people who have personality disorders with those who don't suffer from them can help track down culprit genes more quickly.

Actually, though, this line of research is likely to lead us beyond any simple isolating of single genes. Preliminary work already suggests that rather than single genes or processes, it is entire systems or networks of genes that guide the development of what we perceive as shyness or nervousness or irritability. "I don't think anybody expects that a single gene can explain a lot of the difference between one person and another," says Bruce Cohen, director of the Shervert H. Frazier Research Institute at McLean Hospital in Massachusetts and professor of psychiatry at Harvard Medical School. "Any individual gene tends to explain only a few percent of risk for anything. Genes tend to work together in groups. So what most of us are doing now is studying not individual genes, but groups of genes, and their pathways."

And that means the hunt for the biological drivers of personality will require still deeper, and more wide-ranging, molecular digging. Ultimately, how genes interact with external forces won't be any more or less important than how they interact with each other. It's a task that has only recently become possible as scientists have started to sequence entire genomes and better understand the proteins, enzymes, and other compounds that genes produce. This broader focus may also help to explain some of those conflicting results from the earlier gene-based personality studies. "It may not be that what people reported is wrong," says Cohen about the studies that sometimes found connections between traits and genes and sometimes did not. "Just that the effect of each individual gene is modest."

If personality does turn out to represent a composite of genes, it would make determining the specific contribution of any one of them more difficult—and, at some level, irrelevant—to isolate. Gregor Mendel, who laid out our basic understanding of genetics in a series of experiments with peas, showed that dominant physical traits such as a yellow color were always passed on to succeeding generations if both parents possessed that trait. But personality traits aren't so neatly inherited. "If you have a highly extroverted mom and an introverted dad, their kids won't segregate into extroverted or introverted personalities," says Kendler. "What we see instead is a blending of many genetic variants, each with a small effect. Personality is like height, only worse, with hundreds if not thousands of potential genetic contributors."

However it all shakes out, however wide the net psychologists cast in future studies to snag the sources of personality traits, the genetic piece of personality will still play a critical role. "We've got to study people in more than one way," says Cohen. "What we and others are beginning to do is not just look at one or two things about people but a range of factors—we measure personality traits, look at the genetics associated with normal and abnormal personality states, look at brain imaging and cell studies. That way we can get convergent evidence about what biological factors are related to the expression of a personality." Imaging studies, for one, can identify different regions of brain activity in people with normal and abnormal personality traits, thus pinpointing potential areas of research into how that particular brain functioning develops.

Only by considering the broader, more complex systems that appear to feed personality creation will researchers be able to finally paint an accurate picture of how particular traits emerge. And yet given the complexity of the ways in which genes and environment interact in birthing these traits, it's possible even that advanced work won't be enough. It wouldn't be at all out of the question, for example, for future research to find that similar traits are produced by different processes. That

Rather than single genes or processes, entire systems guide the development of what we call personality.

is, two people may be introverted for two different sets of reasons, with one case perhaps more genetically determined than the other. Still, any gene-based clues are an important step on the long road to fully understanding the biological basis of what makes us who we are.

"We can't begin to have a sense," says Kendler, "of what effect particular brain systems—from brain chemicals such as neurotransmitters to the effects of myelin [which insulates nerve cells]—and any number of other things that influence the functioning of the brain, including genetics, have on personality. We are almost completely ignorant of that."

In the meantime, we can at least agree to change our definition of personality to reflect the burgeoning appreciation of how nature and nurture interact in a dynamic way. "I like the word 'temperament' to refer to a person's predisposition, and 'personality' to refer to the pattern of thinking, feeling, and reacting that emerges in adulthood," says Potash. "So what emerges in adulthood is shaped by environment and experience." Genes, as Potash's definition recognizes, certainly play a crucial role in shaping destiny, but their effects are by no means absolute—and are certainly more complicated than Mendel and the geneticists who followed him could ever have imagined.

What Is She Thinking?

That's the question that empathy helps us answer.
And the earlier we teach it to our kids, the better.

BY MAIA SZALAVITZ

and baby—and, by extension, to understand each other—in a program called Roots of Empathy (ROE).

In a country like the United States that thrives on its citizens' individualism and competitive spirit, empathy is often seen as a luxury or a frill—a personality trait that is admirable, sure, but far from essential. In fact, the ability to understand the minds of others and then to care about what the world looks like from their perspective is a fundamental building block of normal personality, not to mention crucial to attaining happiness and staying healthy. ROE, which is now offered in over 130 American schools as well as throughout Canada and in several other countries, is teaching children to hone that all-important aptitude.

At its core, ROE (along with its sister program, Seeds of Empathy) is a promising anti-bullying and psychological-health curriculum that starts as early as preschool. The nonprofit program is based in part on social neuroscience, a field that has exploded in the past 10 years, offering hundreds of findings on how our brains are built to care and cooperate, not just compete. Once a month, students watch the same loving mom and baby interact on that blanket. Specially trained ROE instructors hold related classes and discussions before and after these visits throughout the course of the school year.

Founded in 1996 in Canada, ROE has taught more than 500,000 children in 10 countries. It reached 65,000 students in some 3,000 classrooms in 2012 alone. To date, 11 independent studies have shown that ROE schools experience "reduced aggression" and "increased prosocial behavior" among participating students.

The goal is to help children learn how to grasp the perspectives and feelings of others. "We love when we get a colicky baby," says the program's founder, Mary Gordon. It gives the mother the opportunity to tell the class how frustrating and annoying it is when she can't figure out what to do to get her baby to stop crying. And that, in turn, gives children insight into a parent's viewpoint—and into how children's behavior can affect adults. In most cases, it is a concept they have never contemplated before.

Like language, empathy is a capacity inherent in the normal human brain. But just as children cannot learn to speak without being spoken to, they cannot become empathetic without being treated kindly. And, as neuroscience research in species ranging from rats to humans has shown, that matters, because being well nurtured as a child is the key to developing an optimally responsive stress system. Unless children are subjected to attentive early care, their brain's stress systems become dysregulated—elevating a risk not just for psychiatric problems, from addictions and anxiety disorders to mood disorders and schizophrenia, but also for physical problems like diabetes, heart attack,

AT A PUBLIC SCHOOL IN Toronto, 25 third- and fourth-graders circle a green blanket, focused intently on the 10-month-old baby with the serious brown eyes who sits at its edge. Baby Stephana, as they call her, crawls toward the center of the blanket, then turns to glance at her mother. "When she looks back like that, we know she's checking in to see if everything's cool," explains one boy. It's just one of the observations he and his classmates have made as they learn to understand the relationship between mother

obesity, and stroke. When they are raised lovingly, the "bonding hormone" oxytocin helps provide a natural brake on stress responses.

Here's how empathy develops naturally. As parents nuzzle and hold their new baby, they learn to respond to the child's unique cries, coos, and smiles. Some research suggests that "mirror neurons" in the brains of both child and adult cause each to react to the other with mimicked responses. So when Baby smiles, Mom does; when Mom smiles, Baby does too. In the course of ordinary, attentive care, a child learns through this process that his mind works like others' minds and that other people have feelings, intentions, and needs, too. Meanwhile, his stress is also calmed by the loving contact.

Very early on, these interactions create the ability to share joy and pain. For instance, infants will cry when other infants cry, with no distinction of whether the pain is their own or someone else's. Soon, though, they move past this mere "emotional contagion." They begin to recognize that because they like to hold Bunny whenever they feel sad, bringing Bunny to a frowning Daddy might make him feel better, too.

Ordinarily, this creates a developmental cascade in which children eventually learn what is called "theory of mind." That is, they come to understand that other people have separate minds, with particular feelings, needs, and motivations that are very likely different from their own. Bunny may soothe Baby, but that doesn't mean it will do the same for Daddy. He, of course, may need a more adult type of pacifier.

The development of empathy has been highlighted in studies of typical toddlers, aged 1 to 2. Research has found, for instance, that young children will, without being prompted, almost always try to help an adult grab a toy that is just out of reach. Further, experiments show, even babies clearly understand intent; they will return a book to a pile if an adult has clumsily knocked it off, but not if the book has been deliberately pushed off by that adult.

Similarly, a sense of fairness develops early, before children are explicitly trained in it by parents and preschool teachers. Infants' surprise can be measured by watching how long they stare at a scene or object. Specifically, studies have shown repeatedly that they look longer at something that violates their expectations. As it turns out, way before they start to drive parents crazy with demands for a slice of cake that is utterly equivalent to those given to their siblings, 18-month-olds look longer when goods are split up unequally among adults who have done equal work than they do when equal work is rewarded with the same amount of crackers and milk.

Oddly, a child's first lie is also an important indication that he understands "theory of mind." If a kid thinks everyone else knows everything that he knows, he will not lie because it makes no sense for him to do so; everyone already knows the truth. It is only when he realizes he knows things that others don't that lying becomes a possibility. Of course, this doesn't mean that children should be instructed in this particular dark art—merely that it's a step on the path to learning that others' minds harbor different ideas.

"Theory of mind" is just one aspect of empathy. Also known as cognitive empathy, it refers exclusively to the ability to intellectually place yourself in someone else's shoes, to reasonably consider what they might be thinking or feeling. The other factor in the equation, maybe an even more significant one, is the capacity to muster concern for how another person feels, beyond simply recognizing that it may be different from what you are feeling. This is emotional empathy.

Problems with both aspects of empathy are evident in many psychological and developmental disorders. Autistic people, for example, tend to develop "theory of mind" more slowly than others do, and they may have difficulty reading facial expressions of emotion. However, they are not impaired in the emotional, or caring, part of empathy once they recognize that someone else is in pain. In fact, they may be so distressed by another's pain that they react by withdrawing rather than helping. In this case, "too much" of one type of empathy actually

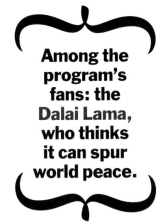

Among the program's fans: the Dalai Lama, who thinks it can spur world peace.

hinders empathetic behavior.

In contrast, people who have antisocial personality disorder are highly skilled at cognitive empathy—that is, they are proficient in understanding the minds of others. What they lack is emotional empathy; they don't care about anyone but themselves. That they can use their cognitive empathy to predict other people's behavior, then not care about causing harm with responses informed by that knowledge, is what makes them potentially dangerous.

Narcissistic personality disorder also involves deficits in emotional empathy, but here it is related to an excessive focus on the self rather than a complete lack of regard for the other. Empathy problems are also seen in schizophrenia and related personality disorders. People with these conditions tend to have delusions about other people's intentions or to see intention in coincidences where it does not exist.

Programs like ROE teach empathy by demonstrating it, building on its origins in early mirroring behavior. This means that children need to understand their own feelings before they can learn to reach out to others. After a child has hurt someone, "we always think we should start with 'How do you think so-and-so felt?'" says Gordon. "But you will be more successful if you start with 'You must have felt very upset.'" The trick, she says, is to "help children describe how they felt, so that the next time this happens, they've got language. Now they can say, 'I'm feeling like I did when I bit Johnny.'"

Once children understand their own feelings, they are closer to being able to understand that Johnny was also hurt and upset by being bitten. Empathy is based on our ability to reflect back others' emotions, and ROE helps children recognize and describe what that reflection actually looks like and means.

Observing infants is a simple and effective way to put that into practice. Their helplessness and cuteness evoke a powerful protective response; tellingly, it is pretty much the opposite of what happens when bullies sense a vulnerability. "Babies are exquisite teachers of empathy because they are theaters of emotion," says Gordon. "They don't hide anything."

Maia Szalavitz is the coauthor, with Bruce D. Perry, M.D., of Born for Love: Why Empathy Is Essential—and Endangered.

The Biology of Joy

Some of us are just born happy, but scientists are starting to understand how the rest of us can learn to lighten up.

By Michael D. Lemonick

YES, YOU CAN IMPROVE YOUR mood! You heard that right! Want to be happier today? All you have to do is train your brain! If this kind of come-on found its way to you through a late-night infomercial or even an anecdotal testimonial from the self-help acolyte in the next cubicle, you'd be right to be suspicious. But Richard Davidson, director of the Laboratory for Affective Neuroscience at the University of Wisconsin, and other like-minded researchers can vouch for it as more than another overhyped, under-proven claim.

To be fair, Davidson didn't believe it himself at first. As the professor of psychology and psychiatry watched a Buddhist monk sink deep into meditation in a campus lab, Davidson was taken aback by the data streaming into his computer from electrodes attached to the monk's skull. But when Davidson double-checked the readings, it quickly became apparent that what he was witnessing was no fluke. As the monk meditated, electrical activity in his left prefrontal cortex increased at a tremendous rate. "We didn't expect to see anything quite that dramatic," Davidson says. If the king of happiness research gets excited about a piece of happiness

research, everyone else needs to take notice.

At the time of his discovery over a decade ago, Davidson was, in fact, looking for just such a link between prefrontal-cortex activity and the sort of bliss that deep meditators experience. Still, seeing that hooked-up brain crackle with activity was unprecedented even for someone with his experience. It made clear, says Davidson, who published his results in *Proceedings of the National Academy of Sciences,* that happiness is more than a vague, ineffable feeling. Rather, it is a measurable physical state—one that can be induced deliberately too.

But wait (as the infomercials shout), there's more! As researchers have gained an understanding of the physical characteristics of a happy brain, they have come to see that those traits have a powerful influence on the rest of the body as well. For example, people who rate in the upper reaches of happiness on psychological tests develop about 50% more antibodies than average in response to flu vaccines; that, says Davidson, "is a very large difference." Happiness and related mental states also appear to reduce the risk—or limit the severity—of many diseases and chronic conditions as well. And not in a small way. According to one Dutch study of elderly patients published around the same time as Davidson's monk work, upbeat mental states reduced an individual's risk of death by half over a nine-year duration. Says Laura Kubzansky, a health psychologist at Harvard's School

of Public Health, in a masterpiece of understatement: "There's clearly some kind of effect."

It makes sense that there would be. Doctors have long known that clinical depression—the extreme opposite of happiness—can worsen heart disease, diabetes, and a host of other illnesses. They know it because the neuro-chemistry of depression, as opposed to that of happiness, has been studied extensively for decades. Until the 1990s, in fact, says Dacher Keltner, a psychologist at the University of California, Berkeley, "90% of emotion research focused on the negative." Which means, Keltner adds, "there still are all of these interesting questions about the positive state."

A growing number of researchers are now focused on answering those questions, including maybe the most fundamental of all: what exactly does happiness look like in a clinical sense? As yet, nobody can say for sure. For now the word "happiness" remains, says Davidson, "a placeholder for a constellation of positive emotional states, from which individuals are typically not motivated to change." Its precise characteristics and boundaries remain to be mapped out.

Then again, as long as subjects can reliably indicate when they're feeling good, researchers can use different brain-imaging technologies—functional magnetic resonance imaging (fMRI), which maps blood flow to active parts of the brain, and electroencephalograms (EEGs), which sense the electrical activity of neuronal circuits—to zero in on the regions of the brain that are involved with such feelings. And their studies consistently point to the left prefrontal cortex.

There is, of course, a chicken-and-egg issue that needs to be raised here: does the prefrontal cortex create the sensation of happiness, or does it merely reflect an emotional state that has been generated elsewhere? Davidson, for one, thinks the answer is a little of both. "We're confident that this part of the brain is a proximal cause of at least certain kinds of happiness," he says. His own research has shown that some humans do seem to be genetically predisposed—with busier prefrontal cortexes—to being happy. Davidson measured left-prefrontal activity in babies less than a year old as they watched their mothers leave the room. Some babies, of course, cried hysterically. The ones who didn't turned out to be the ones with higher left-prefrontal activity. "We were actually able to predict which infants would cry in response to that brief but significant stress," says Davidson. You could say that he confirmed scientifically what parents seem to know instinctively: some babies are more joyful than others.

But at the same time, neuroscientists have learned that the brain is highly plastic, endowed with the ability to rewire itself in response to personal experience—especially before puberty kicks in. One might assume, therefore, that negative experience would destroy a happy personality, and if the bad moments are extreme and

NIRVANA CHECK *As Buddhist monks enter a meditative state, studies showed, activity in the brain's left prefrontal cortex increases.*

frequent enough, that could very well be the case. Davidson learned, though, that in most cases mild to moderate doses of negative experience are actually beneficial. In animal studies in which he compared groups that were moderately stressed when young with those that weren't, he found the former better able to recover from stress as adults. Because deliberately inducing stress in children would be cause for law-enforcement intervention, Davidson based his human research on self-reported stories of stressful childhoods; his conclusions were the same. The reason, he theorizes, is that stressful events provide practice for learning how to bounce back against adversity. Think of it as exercises that strengthen one's happiness muscles.

What distinguishes a left prefrontal cortex that is predisposed to happiness from one that isn't almost certainly has to do with neurotransmitters, the chemicals that ferry signals from one neuron to the next. And while the prefrontal cortex is awash in many of them—serotonin, glutamate, GABA, and more—Davidson has singled out dopamine as the difference maker. Dopamine mediates the transfer of signals associated with positive emotions between the left prefrontal area and the emotional centers in the limbic area at the very core of the brain, and people with a more sensitive version of the receptor that accepts dopamine tend to have better moods. Researchers continue to look into the relationship of dopamine levels to feelings of euphoria and depression.

Dopamine pathways may be especially important in aspects of happiness that are associated with moving toward a goal (monks attaining a meditative state, cigarette smokers lighting up after 24 hours of deprivation). Other neurochemicals, though, may be more central to different kinds of happiness, for example physical pleasure. "We're just beginning to apply a lens to all those parts of the nervous system in which the positive emotions are embodied," says Berkeley psychologist Keltner. "This is really neat territory."

Among those exploring that territory is Brian Knutson, an associate professor of psychology and neuroscience at Stanford, who, like Davidson, uses MRI to monitor the brains of test subjects. The mental mode he studies is anticipation. "When people think of happiness," says Knutson, "they think of feeling good, but a big part of happiness is also looking forward to something." Knutson's research was inspired by the classic work of Ivan Pavlov, who trained dogs to salivate at the sound of a bell by associating it with mealtime.

In place of food, Knutson baited his study with money: a small cash payoff for subjects who won a videogame. "When we looked at their brains just before they got the reward," he says, "we saw this spark that clearly had to do with how positive the idea of making money was." The spark showed up not in the left prefrontal cortex but in what Knutson terms a more primitive section—the nucleus accumbens, located in the subcortex, at the bottom of the brain. The bigger the prize, Knutson found, the stronger the activation. Knutson believes he is looking at the kind of happy feelings we experience as excitement. And although the focus of his work is to understand the neurophysiology of motivation and decision making—how emotion and reason work together as we make choices—it could also be a key to plotting the brain's broader happiness circuitry.

Understanding the neurophysiology of feeling good is one current path of happiness research; another is understanding how positive emotion affects the rest of the body. As in brain studies, the term "happiness" is too broad to inform a rigorous approach, so researchers tend to focus on specific definable qualities. Harvard psychol-

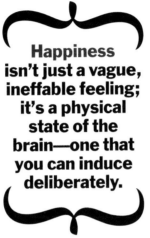

Happiness isn't just a vague, ineffable feeling; it's a physical state of the brain—one that you can induce deliberately.

ogist Kubzansky, for one, has set her sights on optimism. Tracking 1,300 men over 10 years, she found that heart-disease rates among men who called themselves optimistic were half the rates of men who didn't. "It was a much bigger effect than we expected," Kubzansky says. "As big as the difference between smokers and nonsmokers."

She and her team also looked at pulmonary function, because poor pulmonary function is predictive of a range of bad outcomes, including premature mortality, cardiovascular disease, and chronic obstructive pulmonary disease. Again, the optimists were much better off. "I'm an optimist," she deadpans, "but I didn't expect results like this."

Kubzansky has worked with Duke psychologist Laura Richman on another study that targeted hopefulness and curiosity—a couple of mental states that overlap with optimism in some ways. Here, too, the results are telling. "We found them to be protective against hypertension, diabetes, and upper-respiratory infection," Kubzansky says. These protective effects may explain the longevity advantage found in that Dutch study—an advantage, by the way, that persisted even after researchers corrected for diet, education, and other potentially mitigating factors.

Exactly how states of mind affect the body's biochemistry is still far from clear, but researchers are narrowing the areas of their focus. "We can do some good speculation," says Kubzansky, "based on what we know about anxiety and depression, so there are a couple of places to look in terms of neuroendocrine function and immune inflammatory pathways." Here's one clue: in addition to reporting a positive mood when their left prefrontal cortexes are active, subjects in Davidson's experiments have lower levels of cortisol, a hormone produced by the adrenal gland in response to stress. Cortisol is known to depress immune function, so it could follow that optimists may feel less stress than pessimists, which in turn might help them avoid the noxious biochemical cascades that stress triggers.

On a more fundamental level, optimistic, happy types as a rule take better care of themselves than sad sacks do. In yet another series of studies, psychologist Robert Emmons of the University of California, Davis, found confirming evidence that happy people do a better job with personal health maintenance. Emmons randomly assigned 1,000 adults to one of three groups. The first kept daily journals of their moods, including rating them on a scale of 1 to 6. The second group did, too, but also listed the things that annoyed or hassled them throughout their day. The third group kept journals but added an activity that has been repeatedly shown to improve one's sense of satisfaction with life: writing down each day all of the things for which they were grateful.

Despite being assigned randomly, the members of group three not only realized the predicted jump in happiness, says Emmons, but they also spent more time exercising, were more likely to get regular medical checkups, and more routinely took preventive health actions like using sunscreen. Overall, the "gratitude" group did the best job of promoting better health. "They rated themselves as more energetic, more enthusiastic, more alert," Emmons says. Keeping diaries, in short, contributed to both physical and emotional well-being.

Not surprisingly, the advantages enjoyed by the "gratitude" group were greatest when they were stacked up against the responses of group two, the one whose members were asked to focus on life's hassles. "People who are grateful tend to view their body a certain way," says Emmons. "They see life as a gift, health as a gift. They want to take certain measures to preserve it."

The beauty of this finding is that thinking about what you have to be grateful for is a non-weight-bearing workout that anyone can do. More sophisticated methods of

> **Stressful events give us practice at bouncing back from unpleasant emotions, like an exercise to strengthen our happiness muscles.**

manipulating happiness have shown promise as well. For example, although cognitive behavioral therapy and medication are traditionally used to combat depression, they may also be useful in enhancing happiness.

There's a sense among these pioneering researchers that their work fits with a growing understanding among neuroscientists that the human brain, once believed to be fully formed in childhood, keeps developing long afterward—a phenomenon known as neuroplasticity. In essence, the brain can literally rewire itself, forging new connections between neurons as we learn new skills. It's one reason, scientists believe, that learning a new language in middle age can help prevent dementia later in life, and why amputees suffering phantom limb pain can overcome it through biofeedback and other brain retraining. It may well be how monks, yogis, and even readers of the original self-help book, *The Power of Positive Thinking,* achieve their impressive results.

Besides, says Keltner, "once enough findings trickle in to show that positive emotions and happiness make your immune system function better, or help you battle disease, or help you live longer, then you're into fundable territory." And that will end up making all of us a whole lot happier. —*Reported by Dan Cray/ Los Angeles*

The Wisdom of 'Use Your Words'

AH, THE TERRIBLE TWOS: ANOTHER tantrum about putting on socks? As parents bury their heads in their hands, supportive onlookers counsel: "It's just a stage." They're right, mostly. Sometimes, though, the outbursts continue beyond that developmental year, and until recently no one could tell the disheartened parents why. Now, however, researchers have new clues: it may be a child's failure to communicate.

Psychologists at Penn State University published a study in 2012 in the journal *Child Development* that seems to uncover a link between language skills and early emotional growth. Following 120 toddlers from the time they were 18 months until just after they turned 4, the researchers assessed those skills and monitored their development. Periodically, they also observed how the kids handled frustration and anger, presenting the children with a gift-wrapped bag, then forcing them to wait several long, boring minutes before opening it.

The authors went into the study suspecting a link between verbal proficiency and the managing of emotions, assuming that children who acquire language quickly would be better equipped for the task as they would more easily be able to grasp rules, communicate needs calmly, and keep themselves occupied. And the results seem to confirm that hypothesis. "We found that toddlers who have stronger language skills, and whose language skills develop faster, are better at regulating frustration once they're preschool-aged," says Caroline Roben, the lead author of the study, now a postdoctoral researcher at the University of Delaware.

By age 3, while the good communicators were calmly asking how long they had to wait, the less developed kids were fussy and squirmy, ready to throw a fit. By age 4, the good communicators were more adept at keeping themselves amused without toys—by talking, counting, or otherwise playing solitary games. These findings suggest that children with stronger verbal skills—that is, the tools to describe and express emotions and to handle themselves when they begin to feel unhappy—will be better behaved as they enter preschool.

The scientists acknowledged that rather than

setting the stage for angry behavior, slower language development may be more prevalent in youngsters who are more angry to begin with, either because they have trouble focusing on the relevant skill building or because adults may find them unpleasant to deal with and thus avoid interacting with them. It's also possible that an as-yet-unfound developmental process drives both verbal and emotional development.

Given the strength of the findings, however, Roben and her team believe that language affects emotional development. And they hope that leads to remedial interventions that improve emotional learning and thus avert later behavioral problems. In the end, kids who are able to sit still, follow instructions, and play well with others are more likely to get higher grades than their aggressive peers. "Emotion regulation is important for success in school," she says.

Parents, too, can benefit from acknowledging this link, as it can help them focus on advancing their toddlers' development. "When you promote your children's verbal world, you may be doing something to promote their emotional development as well," Roben concludes. Not to mention to escape the terrible twos as fast as you can. —*By Laura Blue*

Nurture

Our notion of environment, once limited primarily to immediate family, has expanded. And though genes paint part of the picture, how they interact with the outside world matters more. So, are you an orchid or a dandelion?

{
Is Childhood Destiny?
The New Science of Siblings
When Parents Play Favorites
The Culture Inside Us
}

Make Yourself at Home

*Where we come from and how we're raised
do matter, but in the puzzle of why we're who
we are, it's just one interlocking piece.*

BY DAVID BJERKLIE

MAGINE AN IMPOSSIBLY LARGE COFFEE-TABLE BOOK. IN THIS enormous tome are snapshots depicting the circumstances into which each of the 350,000 babies born on the planet every day arrives in the world. The sheer variety of conditions—geographic, economic, cultural, social, and familial—would be mind-boggling. Now imagine trying to make an educated guess about what sort of adult each of those kids will become. How will circumstances and experience shape their personalities, their characters, their sense of self?

We have known for a long time that who we are is a combination of the genes we inherit, the environment in which we grow up, and the experiences (good and bad) we encounter along our way. And yet we still often want to see our characteristics mostly as products either of our DNA or of our circumstances. We have an odd mathematical need to apportion credit (or blame). Yes, we all agree, both nature and nurture have important roles to play, but what is the relative importance of each? Is IQ, for example, 90% genetic, or more like 75%? How about depression—or shyness or optimism or pedophilia?

The urge to attach numbers to the influencers of our psychic development may be natural, but many experts wish we'd stop thinking in those terms. To them, it's simply misguided "to try to finely distinguish the relative importance of nature and nurture in the course of human development," in the words of Jack Shonkoff and Deborah Phillips, editors of *From Neurons to Neighborhoods: The Science of Early Childhood Development*. And if the debate seems a little inside baseball, it's not. Too often, say Shonkoff and Phillips, the quantitative approach gives short shrift to the importance of parenting, education, and intervention programs.

MILENA BONIEK/PHOTO ALTO/CORBIS

If not all researchers see things that way, the tide is turning. For every behavioral geneticist who continues to study twins and adoption to calculate "heritability statistics" (estimates of the variability in characteristics caused by DNA differences), there's another who believes heritability estimates have to be seen as dynamic—that is, that a trait that is highly heritable at one age may be less so at another. Besides, just because a trait is highly heritable, they argue, doesn't mean it can't also be malleable. Why wouldn't heritability change over time, if gene activations and environmental influences change throughout our lives too? Scientists may not all be on the same page quite yet, but the ongoing dialogue is bound to yield a richer understanding of development.

No one can be faulted, though, for wanting things to be simpler than they are. Says Andrew Solomon, author of *Far From the Tree,* a human-eye view of the workings of nature, nurture, and identity in the lives of an extraordinary range of people: "I feel that every generation has wanted to find an explanation that made other explanations obsolete. In the '50s and '60s, when psychoanalysis was at its strongest, we wanted to believe that everything was determined by environment. That gave way to a biological determinism that told us, 'It's all in your genes.'"

Today, thankfully, observes Solomon, "it seems we occupy a more intelligent middle ground," one cohabited by genes and environment.

In any event, it's becoming more and more apparent that unraveling the mysteries of human development isn't at all a matter of neatly calculating how much of a trait is attributable to genes and how much to environment. Rather, it comes down to understanding how genes and environment interact with each other to produce that trait. And further, our notion of what constitutes "environment" has expanded. Nurture once referred primarily to parents and immediate-family circumstance, but in fact it is a far larger and more subtle concept. Even within a single family, researchers now recognize, no two siblings experience the same environment, as parenting styles are adapted to the traits and temperaments of each child. That's a natural and mostly good thing, even if there can be negative reinforcement in consistently seeing a child as "the stubborn one" and another as "the peacemaker" or "the drama queen" or whatever.

For a while in the 1990s, some researchers stirred up the nurture pot by making a case for the influence of peers over parents. After all, kids spend most of their

{ By Any Means

*We are shaped by the interactions of our genes
and environment. It is a timeless dance of nature
and nurture, no matter whether it takes place in
circumstances of privilege or poverty.*

At birth, the impact of environment becomes much more obvious, particularly in neurological development. In an infant's first 12 months, the brain nearly triples in size, and over the next few years there are a number of critical or sensitive periods—"windows"—when it demands certain types of input to create or stabilize long-lasting structures. Take the development of the visual system. A baby born with a cataract can look forward to perfect vision in that eye as long as the clouded lens is promptly removed. Wait too long, though, and the baby will become permanently blind, because the brain's visual centers require early sensory stimulus—in this case, light hitting the retina—to maintain what at this young stage are still tentative neural connections.

With very few exceptions, however, the windows of opportunity in the human brain are not in danger of slamming shut so abruptly. There appears to be a series of windows on the way to developing language, for example. While the window for acquiring syntax may close as early as 5 or 6 years of age, the window for adding new words may never close. "It's clear," says University of Southern California neuroscientist Pat Levitt, "that environment plays an incredibly strong role. Studies have found over and over that when you look for the factors that affect language development and complexity of vocabulary, it's what an infant or toddler hears on a daily basis." It's also widely understood, says Levitt, that a child who starts out in a language-poor environment but then moves into a language-rich one will be able to increase his verbal skills. But whether that increase can match what that child could achieve living in a language-rich environment from the first is not certain.

Studies have indeed found that there is a correlation between language exposure and socioeconomic status and level of education. In their 1995 book *Meaningful Differences in the Everyday Experience of Young American Children,* researchers Betty Hart and Todd Risley recounted their experiences with devising language programs for children of families who lived in the housing projects of Kansas City, Kan. Although strides were made, they could not keep pace with the language skills of children born into professional families.

Parents were the critical factor. The psychologists tracked kids in 42 families for more than two years. They found that "despite the considerable range in vocabulary size among the children, 86% to 98% of the words recorded in each child's vocabulary consisted of words also recorded in their parents' vocabularies." By age 3, trends in the amount of talk, vocabulary growth, and style of interaction had been well established. And from the researchers' perspective, that was not encouraging: "While children from different backgrounds typically develop language skills around the same age, the subsequent rate of vocabulary growth is strongly influenced by how much parents talk to their children."

time with other kids. Peer pressure, both positive and negative, suddenly was the 900-pound gorilla in the behavioral lab. Judith Rich Harris, one of the most controversial members of this group, bluntly concluded: "Parenting has been oversold. You have been led to believe that you have more of an influence on your child's personality than you really do." After the commotion died down—there were plenty who decided it was Harris's view that had been oversold—what remained was a slightly uneasy feeling that maybe the influence of environment and experience was too big and broad to ever pin down.

That influence, of course, starts in the womb, where the growth and development of a fetus are affected by what the mother eats and drinks, the air she breathes, how much stress she endures, and the balance of hormones in her body. It goes far beyond contributing to pathologies such as fetal-alcohol syndrome and crack babies. Vulnerabilities to diabetes and heart disease and, most relevant to this discussion, a range of behavioral predispositions also begin in utero.

A child in a family on welfare heard an average of 616 words per hour, a child in a working-class family heard 1,251 words per hour, and a child in a professional family heard 2,153 words per hour. Doing the math to age 4 shows overwhelming differentials between the groups: 13 million words, 26 million words, and 45 million words—a gap of more than 30 million between the welfare and professional families.

The numbers tell a skewed story to some critics, who take issue with what they call the "deficit" perspective. They say these comparisons end up unfairly marginalizing the language and culture of the poor, particularly as families living in poverty are an ethnically, linguistically, and racially diverse group. Still, it is hard to ignore the fact that when children grow up in different linguistic environments, those differences have consequences.

Generally, the advances of pediatrics describe this country's changing concerns over the health of children. As set forth in a 2012 article in the journal *Pediatrics* by Jack Shonkoff, Andrew Garner, and others, the initial focus in the late 19th century, when that medical specialty emerged, was on nutrition, infectious disease, and premature death. By the middle of the next century, after

Neurons and Neighborhoods
There is no way to perfectly distinguish the influence of biology from that of environment in development. What's more, different experiences can produce different outcomes on identical genes.

People Who Need People
Our capacities for resilience develop and change as we develop and change. Some of these capacities are found within the child; others are found within the child's relationships.

vaccines, antibiotics, and other public-health measures had been established as routine care, attention turned to developmental and behavioral problems. By the end of the 20th century, more specific issues such as mood disorders, parental substance abuse, and exposure to violence were getting increased consideration. Today researchers and pediatricians alike devote themselves to ever more complex mental-health concerns.

Consider it the historical record of our understanding of how childhood environments cause lifelong effects. One of the more recent watershed moments in this ongoing arc was the research conducted by Vincent Felitti, Robert Anda, and their team and reported in 1998. The research, which became known as the Adverse Childhood Experiences (ACE) study, was based on surveys of some 13,000 adults about seven categories of childhood experiences: psychological, physical, or sexual abuse; exposure to violence against mother; substance abuse; mental illness; and criminal behavior. The results were striking—individuals who had been subjected to four or more of these adversities had four- to 12-fold increases in risks for alcoholism, drug abuse, depression, and suicide attempts, and two- to fourfold increases in smoking, promiscuity, and sexually transmitted disease, compared with individuals who were exposed to none.

Similarly, researchers have found that children who endure this sort of "toxic stress" are more likely to have trouble maintaining supportive social networks and are at higher risk of school failure, gang membership, unemployment, poverty, homelessness, violent crime, and incarceration, and of becoming single parents and problem gamblers. The bottom line, conclude Shonkoff and Garner, is that "many adult diseases should be viewed as developmental disorders that begin early in life."

Parsing the data further reveals that severe neglect may sometimes have an even worse effect on a child's psychic development than abuse. "People often think of a lack of interaction as just a lack of interaction," says neurologist Levitt. "But, actually, the absence of input is

toxic." Children who have lived through chronic neglect, studies show, are at greater risk for emotional, behavioral, and interpersonal-relationship difficulties later in life, as well as for abnormal physical development and impairment of their immune system.

As compelling as this body of work is, just as genes alone are not destiny, neither is environment. Some children seem to possess a remarkable ability to recover from hardship, neglect, and even abuse. Psychologists call it resilience. What that means, however, goes deeper than that classic coaching chestnut "When the going gets tough, the tough get going." "Some people use the word 'resilient' to describe some sort of intrinsic characteristic or trait," explains Levitt. "But others see resilience as a

characteristic of environments and circumstances that allows one to respond in a positive way to challenging experiences." To what extent it is both is not entirely clear, he says, "but we know it's really, really complicated."

It is also, thankfully, not uncommon. Psychologist Ann Masten of the University of Minnesota's Institute of Child Development calls resilience "ordinary magic." Resilience is all around us, she says. We see it in the many good outcomes among children who have experienced poverty, violence, disaster, or other traumas. Resilience is a process. Various relationships—with caregivers and teachers, friends and family—can promote resilience. Sometimes it even takes a village. Medical anthropologist and psychiatrist Brandon Kohrt has worked in Nepal since 1996, studying with colleagues the lives of Nepalese children, ages 5 to 14, who had been forced by the Maoist rebels to serve as child soldiers in the country's civil war. Not surprisingly, once they returned to their villages these children suffered higher rates of depression and posttraumatic-stress disorder than their contemporaries who weren't conscripted. What Kohrt and his colleagues found, however, was that the particular environment to which these children returned was critical to their postwar mental health. Those who

were stigmatized or ostracized suffered most acutely and persistently, with the nightmarish trauma of battle compounded by social isolation. Children who were reintegrated into the community, on the other hand, were able to reclaim their lives.

But even in resilience, environment always acts in concert with genes. Scientists have known for a while that certain genes put individuals at higher risk for a disorder only if those individuals are exposed to an abusive childhood environment, a case of the wrong place at the wrong time for the wrong gene. But researchers are now learning that some gene variants might be more susceptible to all environmental influences. And that results in a surprising upside: the same gene that makes a child more vulnerable to a harsh environment may also make her more receptive to a supportive one. Researchers call it "differential susceptibility" or "biological sensitivity to context." Less academically, it is known as the orchid and dandelion model. It works like this:

Certain children—orchids—are very sensitive to their environments. If they are forced to endure a lot of toxic stress early in their lives, they will end up doing incredibly poorly by any of a number of measures. Put those same children in highly supportive environments, though, and they often end up maturing into dazzling success stories as adults. Dandelions, meanwhile, are better able to cope under a whole variety of conditions. They aren't usually affected too severely by adverse experiences, but they won't get the most benefit from especially positive ones either. Plot out children on a distribution curve, and dandelions represent the large middle.

This hypothesis may help explain the wide range of outcomes linked to resilience. It may also explain why some genes that appear to confer only disadvantages have nonetheless survived evolution's harsh gaze. Maybe under the right circumstances these same genes actually fuel exceptional outcomes. A gene that can put people at a higher risk for depression may be one of them.

Humans have one of three possible combinations of this particular gene: two versions that put us at low risk of depression; one high-risk and one low-risk; or two high-risk. Under benign circumstances—good in an unspectacular sort of way—researchers see very little dif-

ference in the incidence of depression among the three gene combinations. In the face of severe neglect or abuse, however, only people with the two low-risk variants are protected—that is, their incidence of depression doesn't significantly change. Those with one high-risk and one low-risk variant do show an increase in the incidence of depression, but it is those with the two high-risk variants who are affected most dramatically, with a fivefold increased risk. Then again, put children in an environment that showers them with extra love and support, and it is those with the two high-risk genes who are most likely to bloom like pampered hothouse orchids. The same kids who are likeliest to get walloped by a terrible environment are also likeliest to thrive in an exceptionally nurturing one. And so a gene once seen as high-risk now seems better understood as highly sensitive.

This interplay between a child's environment and his genes is not entirely open-ended, but certainly it continues well past kindergarten. One of the unintended consequences of the earliest research into developmental windows was an emphasis on the period between birth and age 3 that seemed to undervalue the flexibility of development in older children. The first three years of life are indeed a crucial time, but it is a mistake to think that the plaster has set by the time we reach 4. That assumption makes for bad science and even worse public policy, which is why the National Scientific Council on the Developing Child is trying to broaden research parameters and popular perspectives. (Levitt is science director of the council; Shonkoff chairs it.)

Up into the teen years, in fact, changes in the brain are still laying the groundwork for the development of the complex suite of cognitive abilities called executive function. "We call it the 'air-traffic controller' of the brain," says Levitt, "because it is our ability to take in information, filter it in a way that allows us to attend to the most salient details, pull information out of our memory bank, filter out the nonrelevant stuff, and then make a decision, all while keeping our cool if the information we are taking in is emotionally charged." The details of exactly how and when this ability develops are important because executive function plays a role in risk taking, dealing with peer pressure, and all the other things we worry about when we wonder what is going through the minds of our teens.

Executive function is just one more manifestation of the truths that apply throughout childhood: that social, emotional, and cognitive development are inextricably intertwined, and that it is the interaction of genes with environment that produces the myriad facets of who we are and how we behave. That might not be the answer we want sometimes. When it comes to destiny, humans have often felt the pull of extremes. It's all in the stars, or it's all in our hands. Our modern reality is far more complicated. Destiny will never be nearly so black and white as we once imagined.

Past Is Prologue

Children subjected to "toxic stresses" such as homelessness face a higher risk of problems in later life. Experts believe many adult diseases are more accurately viewed as developmental disorders.

The New Science of Siblings

Your parents raised you. Your spouse lives with you.
But it's your brothers and sisters who really shape you.

By Jeffrey Kluger

THE CHRISTORY KIDS
*New Yorkers by birth but with a Swiss mom and a French dad, the Christorys
are growing up multilingual. Big sister Céline tutors Arthur and Emile;
Emile tutors baby sister Elise. Céline has a strong sense of duty; Elise is the
clan clown. Arthur is the most independent—common for a second-born.*

IRST IT WAS OUR PARENTS, PARticularly our mothers. Then it was our genes. Next it was our peers. Researchers have long tried to nail down what shapes us—or, at least, shapes us most. And certainly all of those ideas were good ones, as far as they went.

Once investigators had strip-mined their theories, they still had as many questions as answers. Somewhere a sort of temperamental dark matter was exerting an invisible pull. More and more, scientists think this unexplained force is our siblings.

From the time they are born, brothers and sisters are our collaborators and co-conspirators, our role models and cautionary tales. They are scolds, protectors, goads, tormentors, playmates, counselors, sources of envy, objects of pride. They teach us how to resolve conflicts and how not to; how to conduct friendships and when to walk away from them. Sisters teach brothers about the mysteries of girls; brothers teach sisters about the puzzle of boys. Spouses arrive relatively late in the game; parents eventually go. Who are our true life partners? "Siblings," says family sociologist Katherine Conger of the University of California, Davis, "are with us the whole journey."

It's not that the research has ignored siblings until now. It's just that it has been limited mostly to investigations into birth order. Older sibs are strivers; younger ones, rebels; middle kids, lost souls. The stereotypes were broad, if not entirely untrue, and there the discussion ended.

But that's changing. At research centers here and overseas, investigators have launched a raft of studies into the sibling dynamic, looking at ways brothers and sisters steer one another into, or away from, risky behavior; how they form a protective buffer against family upheaval; how they compete for recognition and come to terms with—or come to blows over—such impossibly charged issues as parental favoritism.

The result is intriguing insights into the people we become. Does the boss who runs a congenial office call on peacemaking skills learned in the family playroom? Does the student struggling with a teacher who plays favorites summon coping skills acquired from dealing with a sister who was Daddy's girl? Do husbands and wives benefit from intergender negotiations once waged with sisters and brothers?

The first thing that strikes contemporary researchers is the amount of time siblings spend in one another's

BLENDING THE SIBS
Carla and LaShawn Stewart of Richton Park, Ill., came to their marriage with three children between them, then had two more. Only five years separate the oldest and youngest, so the kids can be natural companions. The term "stepparent" is not used at home. "We're just parents," says Carla.

MINDS OF THEIR OWN
Sofia Romero (left) shares a warm moment with Alejandra, her twin. The New York City girls have distinct personalities. Alejandra obeys—and sets—rules, and Sofia breaks them. "What amazes me is their temperaments when they were born have more or less remained," says their mom.

University of Pittsburgh, "parents serve the same big-picture role as doctors on grand rounds. Siblings are like the nurses on the ward, there every day." Proximity breeds a lot of intimacy—and a lot of friction.

Laurie Kramer, professor of applied family studies at the University of Illinois at Urbana-Champaign, found that, on average, sibs between 3 and 7 years old engage in some conflict 3.5 times an hour. Kids 2 to 4 top out at 6.3, according to a Canadian study; that's more than one clash every 10 minutes. "Getting along with a sister or brother," Kramer says, "can be a frustrating experience."

But while the fighting is setting parents' hair on end, it's doing some teaching too, specifically about how conflicts, once begun, can be settled. Shaw and his colleagues conducted a years-long study in which they visited the homes of 90 two-year-olds who had at least one sibling, observing the target kids' innate temperaments and their parents' discipline styles. The researchers observed the children again, at 5, this time in a play session with one close-in-age sib. The pairs were shown three toys but given only one to play with. They were told they could move onto another only when both agreed it was time to switch and which one they wanted next.

That, as any parent knows, is a scenario trip-wired for fights—and that's what happened. The next year, researchers went to the children's schools to observe them at play and interview their teachers. Almost universally, the kids who practiced the best conflict-resolution skills at home carried those abilities into the classroom.

Of course, other things could account for what makes some kids battlers in school and others not. But the influence of the most powerful variables—parents and personality—was isolated over the four years of observations. Socioeconomic status, an X factor that bedevils studies like this one, was controlled by selecting families from the same economic stratum. What was left was the interaction of the sibs. "Siblings have a socializing effect on one another," Shaw says. "When you tease out all the other variables, it's play style that makes the difference. You're stuck with sibs. You learn to negotiate day to day."

That permanence makes siblings a valuable rehearsal tool. Adulthood, after all, is practically defined by peer relationships—the workplace, a marriage, the church building committee. As siblings, we may fume, but by nighttime we return to our twin beds in the same shared room. Peace is made when one sib offers a toy or shares a thought or throws a pillow in a mock provocation that releases lingering tension in a burst of roughhousing. Somewhere in there is early training for the employee who sends the e-mail joke that breaks an office silence or the husband who signals that a fight is over by asking his wife where they should go on that fast-approaching vacation. "Sibling relationships are where you learn all this," says developmental psychologist Susan McHale of Penn State. "They are relationships between equals."

Yes and no. Multichild households can also be palace

presence and the power it has to teach social skills. By the time children are 11, according to a Penn State University study published in 1996, they devote about 33% of their free time to siblings—more than with friends, parents, or teachers, or even by themselves. A more recent study found that even teens, who have often begun going their own way, commit at least 10 hours a week to siblings. That's a lot when you consider that after school, sports, dates, and sleep, there aren't many hours left. In Mexican-American homes, where broods are generally bigger, the figure tops 17 hours.

"In general," says psychologist Daniel Shaw of the

courts, with alliances, feuds, grudges, and loyalties. But the touchiest problem in many such situations is favoritism. Parents feel much guilt over the often evident, if rarely admitted, preference they harbor for one child over another. If favorites exist, though, it may be not the parents' fault but evolution's.

The family is a survival unit. Parents agree to care for the kids, the kids agree to pass on the genes, and they all do what they can to make sure no one is eaten by wolves. But the resources that make this arrangement possible are limited. "Economic means, types of jobs, even love and affection are in finite supply," says Penn State psychologist Mark Feinberg. Parents are programmed to notice the child who seems most worthy of the investment. While millennia of socialization have helped us resist this impulse, and we often pour much wealth and energy into caring for the disabled or difficult child, our primal programming still draws us to the pretty, gifted ones.

Conger tested how widespread favoritism is. She visited 384 adolescent sibling pairs and their parents three times over three years, questioning them about their relationships, sense of well-being, and more. She also videotaped them as they worked through sample conflicts. Her conclusion: 65% of mothers and 70% of fathers exhibited a preference for a child—usually the older one. And the kids knew it. "They said, 'Well, it makes sense that they would treat us differently because he's older' or 'we're a boy and a girl,'" Conger says.

Kids often appear to adapt well to the disparity and learn to game the system. "They'll say, 'Why don't you ask Mom if we can go to the mall, because she never says no to you,'" says Conger. But second-tier children may pay a price. "They tend to be sadder and have more self-esteem questions," Conger says. "They feel like they're not as worthy, and they're trying to figure out why."

Think you're not still living the reality show? Think again. It's no accident that employees in the workplace instinctively know which person to send into the corner office with a risky proposal or bad news. And it's no coincidence that the hurt and envy that come when that colleague emerges with the proposal approved and the boss's applause seem so familiar. But what you summon with the feelings you first had long ago is the knowledge you gained then, too—that the smartest strategy is not to compete for approval but to strike a partnership with the favorite and spin the situation to benefit yourself as well. You know it now, but you learned it then.

It's no secret that brothers and sisters emulate one another or that learning flows both up and down the age ladder. Younger sibs mimic the skills and strengths of older ones. Older sibs are prodded to try something new because they don't want to be shown up by a more adventurous younger one. More complex—and in many ways more important—are situations in which siblings don't mirror one another but rather differentiate themselves, a phenomenon psychologists call de-identification.

Alejandra and Sofia Romero, fraternal twins who live in New York City, entered the world at almost the same instant but have gone their own ways ever since, at least in terms of temperament. Alejandra has more tolerance—even a taste—for rules and regimens. Sofia observed this (and her parents observed her observing it), then distinguished herself as the looser, less disciplined one. Sofia is also more garrulous, so Alejandra became more taciturn. "Sofie served as their mouthpiece," says Lisa Dreyer, the girls' mother, "and Alejandra was happy to let her do it."

De-identification helps kids stake out personality turf in the home, but it also has a far more important function: pushing some sibs away from risky behavior. On the whole, siblings pass on dangerous habits in a depressingly predictable way. A girl with an older, pregnant teenage sister is four to six times as likely to become a teen mom too, says Patricia East, a developmental psychologist at the University of California, San Diego. The same pattern holds for substance abuse. According to a paper published in the *Journal of Drug Issues* in 2005, younger siblings whose older sibs drink are twice as likely to pick up the habit too. With smoking, the risk increases fourfold.

Some kids break the mold—and for surprising reasons. East's five-year study found that girls who don't follow older sisters into pregnancy may be drawn not so much to the wisdom of their own choice as to the fact that it's a different one. One teen mom in a family is a drama; two has a been-there-done-that quality. Says East, "She decides her sister's role is teen mom and hers will be high achiever."

Younger sibs may avoid tobacco for much the same reason. In 2003, Joseph Rodgers, a psychologist at the University of Oklahoma, found that while older brothers and sisters who smoke often introduce younger ones to the habit, the closer they are in age the more likely the younger one is to resist. Apparently, proximity in years already made them too similar. One conspicuous way for

SURVIVING A LOSS
Wayne Duvall (center) was 13 when his father died. His older brothers, Gary (left) and Randy, assumed that authority, and Wayne readily accepted it. Although the brothers now live in different cities, their bond remains strong. "In many ways," Wayne says, "they still look out for me."

Sibling Rivalries

For every famous sibling, there are others in the nest who might like the spotlight too. Some get it; some don't. Few find the competition easy.

The Williamses *Serena (left), the youngest of five girls, has 15 Grand Slam singles titles; Venus, 15 months older, holds seven.*

'I've lost patience for being compared to my brothers.'
—Neil Bush

The Bushes *George W. (far right) is the eldest of the six sibs. Jeb (third from left)— a middle child—also entered politics.*

The Mannings *A spinal condition kept Cooper (center), the oldest, out of the NFL where Eli (top) and Peyton star.*

The Kennedys *Father Joe would pit his four sons against one another in sports, education, and politics, leading to both tragedy and triumph.*

The Jacksons *The clan gathered for a backyard photo in 1978. The Jackson 5 were old news by then, but Michael (front row left), the third youngest, would make fresh headlines.*

The Waleses *Harry (left) stands lower in the royal line of succession than big brother William.*

'We are slightly normal. We have a normal side to us.'
—Prince Harry

The Gershwins *George (left) and older brother Ira created more than 700 popular songs.*

The Bonapartes *Second child Napoleon made his seven siblings royals.*

The Marx Brothers *From left: Gummo, Zeppo, Chico, Groucho, and Harpo. Zeppo and Gummo left to become agents.*

CLOCKWISE FROM TOP RIGHT: BILL FRAKES/SPORTS ILLUSTRATED/GETTY IMAGES; NEAL PRESTON/CORBIS; AP; BRIDGEMAN ART LIBRARY; MARK CUTHBERT/UK PRESS/GETTY IMAGES; BETTMANN/CORBIS; JOHN F. KENNEDY PRESIDENTIAL LIBRARY; KEVIN LAMARQUE/REUTERS; GEORGE W. BUSH PRESIDENTIAL LIBRARY

a baby brother to set himself apart is to look at an older sibling's habits, then do the opposite.

Far subtler—and often far sweeter—than the risk-taking modeling that occurs among sibs is the gender modeling that plays out between opposite-sex ones. Brothers and sisters can be fierce de-identifiers. In a study of adolescents in central Pennsylvania, boys unsurprisingly scored higher in such traits as independence and competitiveness, while girls did better in empathetic characteristics like sensitivity and helpfulness. What was less expected was that exposure to an opposite-sex sibling didn't temper gender-linked traits, it accentuated them. Both boys and girls hew closer to gender stereotype and even seek friends who conform to those norms. "It's known as niche picking," says Kimberly Updegraff, a professor of family and human development at Arizona State University, who conducted the study.

But as kids get older, the distance from the other gender must, of necessity, close. Here kids with opposite-sex siblings have a marked advantage. William Ickes, a psychologist at the University of Texas at Arlington, published a study in which he paired male and female students—all of whom had grown up with an opposite-sex sib—and set them to chatting with one another. In general, boys with older sisters or girls with older brothers were less fumbling and kept things flowing more naturally. "The guys who had older sisters had more involving interactions and were liked significantly more by their female acquaintances," says Ickes. "Women with older brothers were more likely to strike up a conversation with the male stranger and to smile at him more than he smiled at her."

If siblings can indeed be such a powerful influence on one another, does that mean all siblings are created at least potentially equal? In particular, what about half-sibs and stepsibs? Do they reap—and confer—the same benefits? Findings are scattered on this, if only because shared or reconstituted families can be so complicated. A dysfunctional home in which parents and siblings hunker down alongside those they're biologically closest to does not lend itself to good sibling ties. Well-blended families, on the other hand, may produce step- or half-siblings who are very close. One of the best studies on this topic has been conducted in Britain among different kinds of nontraditional families. In general, the researchers have found that the intensity of the relationships closely follows the degree of physical relatedness. No hard rules have emerged, but the more genes you share, the more deeply invested you tend to grow. "Biological siblings just get into it more," says Thomas O'Connor, a professor of psychiatry at the University of

Rochester Medical Center. "They are warmer and also more conflicted."

One great gift of the sibling tie is that while warmth grows over time, conflicts fade. Even the fiercest wars leave little lasting damage. Indeed, siblings who battled a lot as kids may be closer as adults—and more emotionally skilled too, clearly recalling what their long-ago fights were about and what they learned from them. "I'm sensitized to the fact that it's important to listen to others," a respondent wrote in a study conducted in Britain. "People get over their anger, and people who disagree are not terrible," wrote another. Even those with troubled sibs got something valuable: patience, acceptance, and cautionary lessons. "[You] cannot change others," wrote one. "[But] I wasn't going to be like that."

Full-blown childhood crises may forge even stronger links. The death of a parent blows some families to bits. But when older sibs step in to help raise younger ones, the dual role of contemporary and caretaker can lay a foundation for an indestructible closeness. Wayne Duvall, a New York actor and the youngest of three brothers, was 13 when his father died. His brothers, who had let him get away with all manner of mischief when both parents were around, intuitively knew that they no longer had that luxury. "I vividly remember them leaning down to me and saying, 'The party's over,'" Duvall says. "My brothers are my best friends now."

Such powerful connections become even more important as the inevitable illnesses or widowhood lead us to lean on those we've known longest. Even siblings who drift apart tend to drift back together as they age. "The relationship is especially strong between sisters," who are more likely to be predeceased by spouses than brothers are, says Judy Dunn, a developmental psychologist at London's Kings College. "When asked what contributes to the importance of the relationship now, they say shared early experiences, which cast a long shadow for all of us."

That shadow—like all shadows—is created by light. Siblings, by any measure, are one of nature's better brainstorms, and the new work on how they make us who we are is one of science's. But outside the lab, we see our circumstances in a more primal way. In a world that's too big, too scary, and too often too lonely, there's nothing like having a band of brothers—and sisters—to venture out with you.

6.3 times an hour, siblings between 2 and 4 years old engage in conflict.

36% of people polled say they've become closer to their siblings with age.

Adapted from The Sibling Effect: What the Bonds Among Brothers and Sisters Reveal About Us, by Jeffrey Kluger, with permission from Riverhead Books, a member of the Penguin Group (USA). Copyright 2011.

Playing Favorites

*Never mind what your parents told you. They had a favorite child—
and if you have kids, so do you. Why it is hard-wired into all of us.*

By Jeffrey Kluger

MOM AND DAD WILL say it earnestly: "I do not have a favorite child." But in an overwhelming share of cases, they're lying through their teeth. From clan to clan, culture to culture, one of the worst-kept secrets is that parents *do* have a preferred son or daughter. And the rules for acknowledging it are the same everywhere. Favored kids recognize their status and keep quiet— the better to preserve the sweet deal they've got and to keep siblings off their back. Unfavored kids howl about it like wounded cats. And to the death, parents deny it all.

The stonewalling is understandable. Parents want to avoid the pain their candor could cause, not to mention the pitiless—even furious—broadsides of public opinion. When a mother of two wrote a post on Babble.com headlined "I Think I Love My Son Just a Little Bit More," she was, predictably, torched. "Please work on your issues lady!" was a typical reply. "I feel absolutely horrible for your daughter!" read another. But then there was this: "I completely understand. I too feel this way."

Most parents do. In one often-cited study, 65% of mothers and 70% of fathers showed a preference for one child. And those figures are likely lowballs; subjects tend to mask unacceptable preferences when a researcher is watching.

If scientists can't see through the ploy, kids do. From the moment they're born, brothers and sisters know they are vying for the precious resource of parental attention and try to establish the identity that will catch Mom's or Dad's eye. I'm the smart one! I'm the funny one!

Who will win the sweepstakes is tough to predict. The father-son bond is iconic. Or is it the father-daughter one? A mother understands her daughters—unless they are mysteries to her. It's equally hard to predict favoritism's fallout. Being the favorite may boost confidence, but it can also foster a sense of entitlement. Unfavored children may grow up wondering if they're unworthy of love, or be pushed to forge stronger outside relationships. And there's no telling how differing treatment will shape relationships among the kids.

"My mom didn't like my older sister and did like me," says Roseann Henry, an editor and mother of two girls. "Everyone figured I had it great, except my sister tortured me pretty much all the time. And really, what affects a kid's life more, the approval of a parent or the torment of a sister?"

If the parental habit of ranking children can cause such pain, why has it become an ensconced part of human nature? As with so much else, blame the narcissistic survival instinct to replicate through future generations. It impels Mom and Dad to tilt toward the healthiest offspring.

The most conspicuous sign of fitness, of course, is physical appearance, and parents have a connoisseur's eye. I was the second of an all-boy foursome, and by almost any measure the third in line, gorgeous Garry, should have been the favorite.

And yet he wasn't. For my father it was Steve, the oldest. Firstborns are often favorites, for a reason businesses get: the rule of sunk costs. The more effort you've put into developing a product, the more committed you are to seeing it succeed. "There's a kind of resource capital parents pour into firstborns," says Ben Dattner, a business consultant and organizational psychologist at New York University. "They build up equity in them."

To be fair, the equity does pay off. Historically, the oldest have been tallest and strongest, because early on, they don't have to share food stores with other kids. A Norwegian study also found a three-point IQ advantage in firstborns, partly a result of being the parents' sole focus for a while. These benefits accrue like compounding interest. A small IQ advantage may yield a similar edge in SAT scores, which may tip a firstborn

off the MIT waiting list and into the entering class.

My mother's pet was Bruce, the youngest. In a way, that was my father's doing too. He wasn't excited about having a fourth child so soon after the third and expressed his displeasure in unsubtle ways. My mother countered his negative bias with a fiercely protective positive one.

Favoring a vulnerable child is counterintuitive, in survival terms. My father's hostility should have doomed my baby brother in my mother's eyes too. In the animal kingdom, a child who's being ill treated by one parent has hurdles to overcome just getting out of childhood in one piece. Best for a mom with years of child rearing ahead to cut her losses early.

Compassion clearly guided my mother. But so did other practices we share with nonhuman species. Consider the coot. Unlike other birds, coots don't pour most of their parenting efforts into their strongest chicks but rather spread care around, hoping to maximize the number of offspring that survive. This means actually favoring the weakest, the ones that need the help.

Bruce was cashing in on another important driver of favoritism as well: gender. Parents with cross-gender preferences—the dad who's helpless in the face of a daughter's charms, the mom who adores her prince of an eldest son—are the most obvious examples. But such patterns are not always what they seem.

Studies have found that what parents seem to value most in their opposite-sex kids are the traits associated with their own sex: the sensitive mom with the poetic son, the businessman dad with the M.B.A.

Fully 65% of mothers and 70% of fathers exhibit a preference for one child. Many of the others just hide it well.

daughter. Reproductive narcissism again rears its head. If kids can't look like you, they can at least act like you. And if sometimes children come by those traits innately, they may also adopt them tactically, to court a little extra love. "Siblings are devilishly clever," says family expert Frank Sulloway of the University of California, Berkeley. "They're constantly trying to fine-tune their niche to squeeze the maximum benefits out of their parents."

Gender's power may be magnified in three-child families. As a rule, first- and last-borns have the best shots at being the favorite. In all-boy or all-girl broods this is especially so, as the middle one stands out neither by birth order nor by sex. That's the case too when the sequence is boy-boy-girl or boy-girl-girl. Shifting the order, though—to boy-girl-boy or girl-boy-girl—can change everything. "If you have a child who is different for any reason, especially being an only girl or boy," says psychologist and sibling expert Catherine Salmon of the University of Redlands in California, "that child is going to get extra attention and investment."

Once favoritism is established, it's hard to break. Still, it can fluctuate, depending on what are known as family domains. There's what happens inside the home and what happens outside it. The ex-jock father who favors his athletic son may be driven to distraction by the boy's restless energy when it is time to sit and talk. So when Dad is looking for quiet bonding, he turns to his daughter. Over the course of a childhood, the son may come out on top, but the daughter will get enough attention to render the disparity insignificant. According to Corinna Jenkins Tucker, an associate professor of family studies at the University of New Hampshire, "The problem is when a child isn't favored in any area at all." Psychologists—to say nothing of parents—rightly wonder what the long-term result of that problem may be. Can you grow up staring at the crowned prince or princess across the room and not develop psychic scars?

The experts' advice to parents is simple: if you have a favorite, keep it to yourself. Even if the kids see through the ruse, the fact that you are trying to maintain it can help them preserve the pretext too. The effort it takes is, in its way, an act of love toward the unfavored child. "Perception is the key," says Shawn Whiteman, an associate professor of family studies at Purdue University.

A powerful one at that. Psychologist Victoria Bedford, professor emerita at the University of Indianapolis, has looked at the impact of what's known as LFS (least-favored status) on children's self-esteem, socialization, and relationships with other family members. "My main conclusion was how horrible favoritism is on siblings," she says flatly.

Clare Stocker, a research professor in developmental psychology at the University of Denver, studied sibling pairs over time and found that kids who felt less loved than their siblings were more likely to develop anxiety, low self-esteem, and depression. Some subjects exhibited behavioral problems that led parents to crack down on them, which only widened the gap between the treatment Mom and Dad meted out to each.

The damage that can be done to the unfavored child is easy to understand. Harder to fathom are the ways a best-loved son or daughter can suffer. They go deeper than the resentment a first-tier child fields from sibs. The biggest risk may be that when early life is punctuated by the huzzahs of parents, you may be unprepared for a larger society in which you're one of many, where the charms Mom and Dad saw are invisible to everyone else.

The theme of the prince struggling with adulthood pervades drama and history. Arthur Miller's *Death of a Salesman* is about more than the tragic fall of Willy Loman. It's also about the crisis of his sons—particularly the older Biff—who grew up on a diet of paternal praise. When Biff learns that the real world expects laurels to be earned, he, not wrongly, blames Willy for making him ill equipped to thrive. Favored kids may not learn the lesson Biff did, but it almost always awaits them. As family expert Judith Rich Harris once said, status conferred at home "doesn't travel well."

Favored sibs have other burdens too—among them a sense of guilt. It's hard not to feel for brothers and sisters who are denied the preferential treatment you receive. Then again, even the most blatant favoritism is easier to take when there's a defensible reason for it. Maybe the most extreme example is when a child has special needs. Children with Down syndrome or autism need a particular kind of parental attention. That time and focus are subtracted from the amount assigned to the other kids. Still, even the most tolerant siblings eventually feel they deserve their full share of care. Explaining why the differential treatment is important is the best way to limit resentment. "Research suggests that differential treatment may have no negative effects when children understand why," says Whiteman.

One of the best things about favoritism conflicts is that they usually fade. Usually, though, is not always; slights may never be forgotten. Life issues such as who becomes the caretaker of aged parents can be occasions to relitigate old grievances.

For the rest of us, the truth ceases to sting. My middle-aged brothers and I—Bruce included—still try to coax our septuagenarian mom to concede the obvious. One day she and I were recalling a long-ago school talent show that, like so many others, starred Bruce. She beamed throughout his performance. I glowered.

"Why were you mad?" Mom asked. "Because he was my favorite?"

"Aha!" I cried. "You admit it!"

She looked at me innocently and blinked. "I have no idea what you're talking about," she said.

OBVIOUSLY I love both of my daughters exactly the same amount.

Even if your kids know you have a favorite, the effort it takes to pretend it's not so can itself be an act of love.

Adapted from The Sibling Effect *by Jeffrey Kluger, with permission from Riverhead Books.*

The Culture Inside Us

Some countries and societies really do produce cheerier citizens (keep smiling, Bhutan!). Can where you come from make you sadder or more anxious too?

By Bryan Walsh

ALL ITS COLORFUL MONKS and medieval archers can make the Himalayan nation of Bhutan seem like a fairy-tale kingdom, one where everyone gets to live happily ever after. It's not—locals face a harsh economic reality. A per capita income of less than $2,500 makes Bhutan one of the world's poorer countries. Yet while life there can be a struggle, the Buddhist nation regularly punches well above its weight in global rankings of happiness and life satisfaction.

Why are the Bhutanese so disproportionately cheerful? Maybe because they want to be. In 1972, the fourth king of Bhutan declared gross national happiness (GNH) to matter more than gross domestic product (GDP), and his successors have focused as much on increasing the nation's collective happiness as on improving its economy. They've built Bhutan around the four pillars of GNH: sustainable economic development, yes, but also environmental conservation, cultural preservation, and good governance. It's the pursuit of happiness, Bhutanese style, and it has caught the attention of scholars and policymakers who are trying to foster more sustainable prosperity around the world. "Bhutan is onto something," says Claire Van der

Vaeren, the head of U.N. operations there.

Bhutan may be the poster child of a growing academic interest in measuring happiness on a national scale, but it is a discipline hampered by an unavoidable truth: different cultures express emotions—happiness, sadness, anxiety—differently. In fact, Eastern and Western cultures have very different ideas of the self itself, and that affects how their people perceive the world and act in it.

Surveys show that happiness—like natural resources—isn't evenly distributed around the globe. Richer countries (no surprise) tend to be happier; those enduring longstanding civil strife tend to be less so. And yet the relationship between happiness and wealth isn't seamless. Research by Ruut Veenhoven, a professor of happiness studies at Erasmus University in the Dutch city of Rotterdam, indeed shows that people living in countries where the average income is less than $10,000 a year are likely to skew unhappy. However, once countries pass that $10,000 threshold—the point at which basic needs are being met—money and happiness decouple.

Take Japan. Between the end of World War II and the bursting of the country's economic bubble in the early 1990s, every trend line concerning money and goods soared straight upward. From 1960 to the late 1980s, the country's GDP quintupled—the fastest growth rate in world history to that point. In just a couple of generations, Japan advanced from the brink of starvation to one of the globe's most prosperous nations. During that period, though, measurements of its citizenry's well-being barely moved. Similarly, in the U.S. of the 1950s, about one-third of the people described themselves as "very happy," and the proportion remains about the same today, even though per capita income has more than doubled. This plateauing also exists in other now-rich nations that rallied out of the wreckage of war, like Germany and South Korea. The phenomenon—that beyond a certain income level, happiness ceases to rise—is known as the Easterlin paradox, named for the American economist Richard Easterlin, who discovered it. Apparently, money can buy happiness—just not a lot.

Even rich nations, though, report a wide range of well-being levels, and how those rankings shake out geographically can help identify the deeper sources of happiness. Lesson No. 1: it's good to be Scandinavian. In Veenhoven's most recent World Database of Happiness, five of the top 10 nations are in Scandinavia, with those melancholy Danes, in an upset, leading the group. Swedes, Finns, and Norwegians are well off, and that certainly bolsters their countries' cause. Still, they're not significantly richer than their sadder European neighbors, and, anyway, we know income is only part of the story. Scandinavia's generous social spending and vacation policies don't hurt, but what ultimately distinguishes the region is its broadly egalitarian societies: these countries have much lower levels of economic inequality than similarly well-off nations like the U.S., let alone desperately unequal ones like China or South Africa.

As it turns out, equality is an important contributor to happiness. Social comparisons may be odious, but humans are hard-wired for making them. Our well-being is relative—when we see neighbors doing better than we are, our self-image and happiness suffer. That's a much rarer scenario in an egalitarian culture. Equality also encourages social trust, which, according to the U.N.'s

NO TIME FOR HAPPINESS *Long work hours and elevated stress levels may contribute to low rankings of well-being in Japan. Could there be a genetic component as well?*

2012 World Happiness Report, also promotes happiness.

That report contained a study which found that recent increases in inequality in both the U.S. and Europe have—other things being equal—produced reductions in happiness. The effect is more pronounced in Europe, probably because of ideological differences: some 70% of Americans believe that the poor have a chance of escaping poverty; only 40% of Europeans do.

But here's where things get really interesting. A quick peek at Veenhoven's rankings reveals that the country at the top isn't rich at all. It's Costa Rica, the Central American nation with a per capita income of $11,900. While that is above Veenhoven's cutoff, it isn't by much. And Costa Rica isn't the only Latin American nation that reports lopsided levels of happiness. Five others crack Veenhoven's top 20, including Panama, Mexico, and Colombia. It's a curious enough occurrence that researchers have given it a name: "the Latino bonus."

As Eric Weiner explains in his book *The Geography of Bliss,* Latinos derive much of their happiness from their sprawling but close-knit families, a living situation that is less common in the developed world. Surveys also show that Latin Americans tend to be more optimistic than citizens of other regions. A Gallup poll released at the end of 2012 found that eight of the 10 most positive nations (according to self reports) are in Latin America. Optimism may not be a necessary condition for happiness, but it definitely seems to help, even when incomes aren't particularly high. It's an observation that

SUPPORT NETWORK *The people of Costa Rica (like Karina Gámez Bran, surrounded by members of her extended family) are, according to one measure, the world's happiest.*

argues loudly for a cultural element to happiness.

Meanwhile, many rich Asian nations like Japan and South Korea do not rank high on Veenhoven's scale. The discrepancy may be the result, in part, of the punishing work hours and elevated stress levels often reported in East Asian countries. But stress is a part of life in every developed nation, and by some measures—economic inequality, social trust—East Asian nations actually score better than comparably wealthy countries in North America and Europe.

Could a genetic component be at work? Jan-Emmanuel De Neve, a researcher at University College London, has looked at a gene that encodes a protein that transports serotonin, which is believed to play a role in mood

regulation. The gene has two variants, one short and one long, with the long one producing more of the transporter-protein molecules. De Neve found that subjects with one long variant were 8% more likely to rate themselves as satisfied than those with none, while subjects with two long variants were 17% more likely. Interestingly, Asian-Americans on average had fewer long variants than white Americans or African-Americans.

Whether or not this genetic connection holds up, there seems to be something relevant in the way East Asians define happiness. For most Europeans and North Americans, happiness is linked predominantly to individual fulfillment. How our pursuit of that goal affects other people may tweak our own joy, but it won't define it. That's not the case in East Asian societies, which have long embraced a collectivist ideal. There a person's sense of well-being comes not from realizing individual objectives but from being a harmonious member of a group, be it the family, the workplace, or the country. A 1997 study by the Taiwanese scholar Lao Lu found that Chinese students, but not British ones, reported that social integration and human interconnectedness led to happiness. "Contributing to the society is the ultimate happiness, whereas hedonistic striving for happiness is regarded as unworthy and even shameful," wrote Lu. "For Confucians, happiness is no longer a set of living conditions; it is the psychological state or spiritual world of a living individual." When East Asians report lower levels of happiness than their Western counterparts, they may be answering another question altogether.

Such a fundamental cultural contrast is likely to affect more than happiness. Turning back to those gene studies, a 2009 paper found a positive correlation among higher levels of the short variant, mood disorders, and collectivist political systems. Could it be that East Asians are more prone to anxiety and, because of that, devise political systems that emphasize social harmony to manage it?

In any case, these cultural differences are why any effort to build a national development system around happiness will continue to be challenging. (It will always be easier to calculate GDP than GNH.) But they shouldn't paint the efforts of happiness researchers like Veenhoven or countries like Bhutan as a waste. More countries will be joining the developed world, and as they do, the level of satisfaction we all enjoy by growing richer is going to top out. Policymakers will want to be ready with other ways to keep their citizenries content. Until then, if you're feeling blue, think about hopping the next flight to Copenhagen or Costa Rica.

Types

There are many kinds of people—16 to be exact. Don't be mad; it's true. (Be careful—you don't want that temper to define you.) The good news is, decent bosses can be made from lots of them.

{
WHAT TYPE ARE YOU?
ANGER: A LOVE STORY
WHAT MAKES A LEADER
}

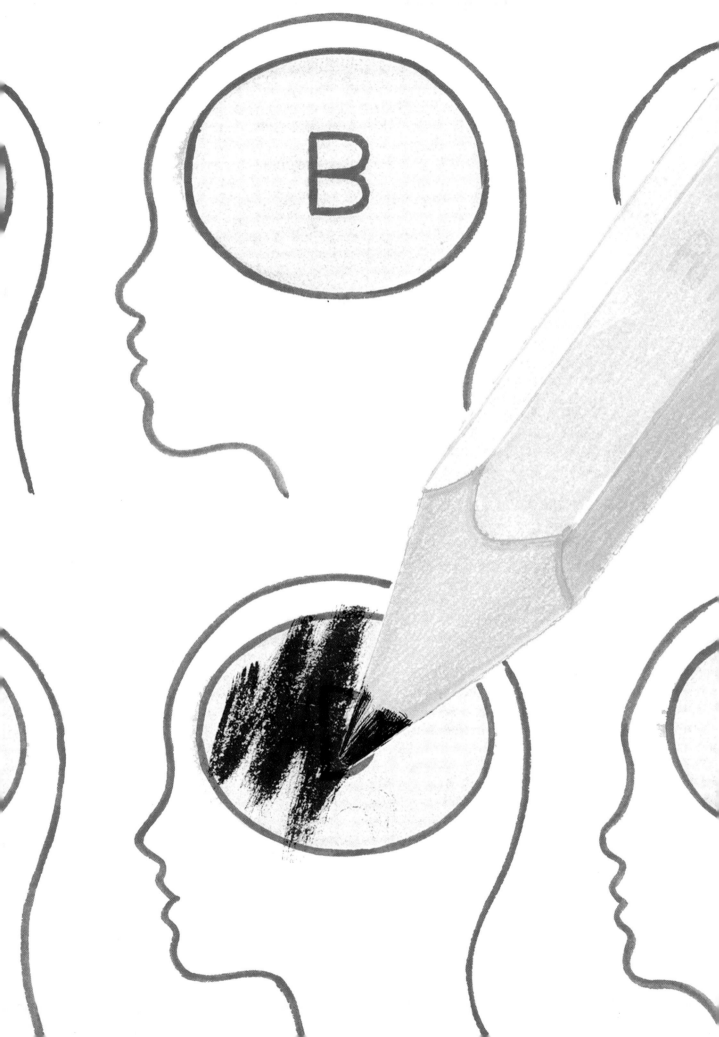

What Type Are You?

*Personality tests are more popular than ever,
as people yearn to answer that question. But it's not so
easy to put the human psyche in a box.*

BY BRYAN WALSH

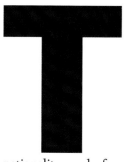HERE SEEM TO BE INFINITE ways to categorize yourself on Internet dating sites, as I discovered when I took a spin through one recently. Some may even invite accurate responses. You have favorite films and favorite foods, favorite hobbies and favorite pastimes. You have age and religion, occupation and nationality—and, of course, height and weight. You can be looking for something temporary or hoping to settle down for life—or at least for a while.

But in addition to all the vital data, I noticed something else popping up in profiles throughout the site: four-letter codes. ISFJ. ENFP. INTJ. If those abbreviations look familiar, you've probably taken the Myers-Briggs Type Indicator test, as more than 2 million people do every year. And that means you know that each quartet represents one of the 16 Myers-Briggs categories. The test was developed during World War II to help women entering the workforce find jobs that suited them, but today the questionnaires—or some bastardized version of them, given the death grip the Myers & Briggs Foundation keeps on the original—have become well-trodden paths to an easily comprehended accounting of who we are. By turning personality into something that can be quantified and qualified, Myers-Briggs has given users, not to mention your potential soulmate, a handy tag with which to advertise themselves.

This sort of personality taxonomy seems the perfect product of a Facebook age populated with those only too happy to define themselves by their likes. But in fact humans have been trying to categorize their psyches for centuries. Traditional Chinese medicine had its six temperaments; medieval European doctors sorted the masses into four humors. (Combative personality? Must be too much phlegm.) Personality typing as we now know it can be traced back to the 19th century, specifically to the work of the British scientist Sir Francis Galton. Today Galton is best remembered—if less than fondly—as the father of eugenics and an early proponent of the harsh theories of social Darwinism (he was a first cousin of Charles). But give him some credit; he happened to be ahead of the curve in theorizing about emotional reactivity, the idea that we all respond to the world in different ways and that those differences can be quantified. Such thinking brought personality "within the fold of scientifically studiable entities," writes the psychologist Daniel Nettle in his book *Personality*.

There's a reason that Galton saw emotional reactivity as an indicator of personality types. Reactivity is really just another way of distinguishing introverts (people who are quieter, more solitary, and more easily emotion-

These can seem like tests for a Facebook age in which we define ourselves by our likes. But humans have been trying to categorize their psyches for centuries.

ally aroused) from extroverts (louder, more outgoing, and less reactive). And the introvert/extrovert spectrum forms the spine of most of today's personality taxonomies. We humans are different in uncountable ways, but nearly all of us can be categorized by whether we more readily choose to turn toward the world or toward ourselves. In fact, these particular inclinations run so deep that they can be reliably identified before a baby is out of diapers. In a landmark 1989 study, the Harvard developmental psychologist Jerome Kagan exposed a group of 500 four-month-old infants to startling scenarios, including popping balloons and brightly colored mobiles. About 20% of the infants reacted intensely, crying and pumping their arms. About 40% stayed quiet. The remaining 40% fell somewhere in between.

Kagan predicted that the infants who had the most extreme responses—the high-reactives—would be more likely to be introverted as adolescents, while the low-reactives would more likely be extroverted. That may seem backward—aren't retiring introverts, by definition, low-reactives?—but it's not. Actually, introverts have a lower threshold for stimulation than others, and that is why they tend to manage their exposure to highly stimulating social situations. Meanwhile, because extroverts can manage more clamor without being overwhelmed, they gravitate toward environments that provide it.

When Kagan brought his subjects back to the lab once they'd grown older, the high-reactive infants had indeed become more introverted. "There's a strong footprint on temperament that you see early in life," says Dr. Carl Schwartz, a psychiatrist at Massachusetts General Hospital in Boston and a former student of Kagan's. "It's not deterministic, but if you're a highly reactive baby, you're less likely to become a bond trader or Bill Clinton."

The elemental nature of the introvert/extrovert continuum—that is, that you can see it at work even among preverbal infants—is what makes it a core part of personality typing. Let's take another look at the Myers-Briggs test. Devised in the 1940s by a Philadelphia housewife named Isabel Myers with some help from her mother, Katherine Briggs, it was one of the first psychological exams that focused not on diagnosing pathology but on categorizing normal. Drawing from the noted psychologist Carl Jung's typological models, the self-schooled Myers and Briggs developed a detailed questionnaire designed to ferret out the particular preferences that make

up a person's worldview, and thus his or her personality. The questions are straightforward: "Do you usually: A) share your feelings freely, or B) keep those feelings to yourself?" Scorers sort respondents into one or the other half of four dichotomies, or pairs of opposing traits. The introvert/extrovert split figures most prominently, but Myers-Briggs also divides between sensing and intuition, thinking and feeling, judging and perceiving.

The sensing-and-intuition pair represents the yin and yang of how we observe and process new information. Individuals who prefer sensing seek information that is tangible and concrete—think engineers and computer scientists—while those who tilt toward intuition go for theory and abstract thought. Thinking and feeling cover our decision-making faculties. Thinkers are detached and logical, marching from one data point to the next, while feelers are less bound to rigid rules of thought or behavior. Judging and perceiving build off the previous two sets, distinguishing those people who are more likely to think or feel (judging) from those who sense or intuit (perceiving) in their dealings with the world at large. Lay it all out and it adds up to 16 different combinations, each expressed by a nifty four-letter code.

With its black-and-white simplicity, it shouldn't be surprising that the Myers-Briggs system was embraced early on by the corporate world; even today it remains a tool of corporate headhunters and human-resources divisions alike. But like most personality tests, it is far from perfect. Relying on self-reported answers, it puts itself at the mercy of subjects who might shift their responses to mimic more socially acceptable results. (The U.S. is a nation of extroverts, or so it appears from all those reality TV shows, and it's not hard to imagine a natural introvert stretching the truth to attract someone in the market for a more outgoing type.) In fact, even in the best of circumstances, the test's accuracy is debatable. A 1991 National Academy of Sciences review found that only the introvert/extrovert scale showed statistically strong validity. The results of the other dichotomies were nebulous at best. For all its popularity—and maybe *because* of its popularity—the Myers-Briggs test may be better suited for an online dating site than for anything that requires more serious science.

But though Myers-Briggs has its problems, it is still a useful window into how personality theory can work. And what we see through the pane is this: personality is

A Spectrum of Personalities

Extrapolated from the typological studies of Carl Jung, the Myers-Briggs personality types have emerged as the best, or at least the most widely used, way to categorize the varied human mind. Here's how the 16 Myers-Briggs types break down—and some famous people who represent each.

E: Extroversion
I: Introversion
S: Sensing
N: Intuition
T: Thinking
F: Feeling
J: Judging
P: Perceiving

ISTJ: The Duty Fulfiller
DWIGHT EISENHOWER
Responsible, quiet, interested in security. Works steadily toward identified goals.

ISTP: The Mechanic
STEVE JOBS
Introvert who wants to find out how systems work. Adept at keeping an open mind to new ideas and information.

ISFJ: The Defender
MOTHER TERESA
Desires order and harmony; gets pleasure from taking care of people. Can be quiet, introverted, and slow to connect.

ESTP: The Doer
JOHN F. KENNEDY
An active, hands-on learner and a natural-born leader. Good at selling ideas and self to other people.

ESTJ: The Guardian
HENRY FORD
A practical, realistic person with a natural head for business or mechanics. Can often excel as an administrator.

ESFP: The Entertainer
RICHARD BRANSON
Active, creative type who takes a hands-on approach to learning and doing. Despises routines, and tends to be play-minded.

ENTJ: The Executive
MARGARET THATCHER
Competent and analytical; typically has a strong will and often gets own way. Can appear arrogant and insensitive to others.

INFJ: The Protector
CARL JUNG
Highly conscientious and sensitive, and often a quiet leader. Very perceptive about the emotions of others.

ESFJ: The Caregiver
DESMOND TUTU
Projects authentic warmth and is skilled at bringing out the best in people. Values tradition and seeks others' approval.

ENTP: The Visionary
RICHARD FEYNMAN
Clever and cerebral, excellent at generating original ideas but not so much in following through. Can have a perverse sense of humor.

INFP: The Dreamer
VIRGINIA WOOLF
Imaginative and spiritual. Finds pleasure in the act of creation and focuses most energy on a personal world within.

ENFP: The Inspirer
WALT DISNEY
Highly energetic driver of change who craves novelty and expressions of strong emotion. Natural leader, especially in the arts.

ENFJ: The Giver
NELSON MANDELA
Outgoing, adept at building large numbers of friendships. Sensitive to others' feelings and good at showing care.

ISFP: The Artist
BOB DYLAN
Sensitive and quiet, and at the same time intensely conscious of the larger physical and emotional world.

INTJ: The Scientist
STEPHEN HAWKING
Analytical introvert who prefers solitary work, often favoring theory over practice. Intelligent, pragmatic, and logical.

INTP: The Thinker
ALBERT EINSTEIN
Quiet, thoughtful, and analytical, prizes autonomy, and is drawn to working alone. Often a good writer.

The assumption that personality remains unchanged over the years doesn't jibe with evidence suggesting that as we age, so does our outlook.

complicated. No one is always one fixed type. Rather, we are a mix and mash of traits, our own unique recipe made from the various ingredients that all of us share.

This is the guiding principle of the Five Factor Model, one of the broadest and most descriptive personality-type systems. Initially advanced by the psychologists Ernest Tupes and Raymond Christal in the 1960s, this taxonomy uses five overarching domains that contain most known traits, and thus represent a sort of skeleton of personality. The domains may sound familiar: extroversion, openness to experience, conscientiousness, agreeableness, and neuroticism. Extroversion characterizes where an individual falls on the introvert/extrovert scale. Openness to experience gauges intellectual curiosity and attention to beauty and novelty. Conscientiousness measures self-discipline and dutifulness. Agreeableness reflects the inclination—or lack thereof—toward compassion and cooperation. And neuroticism rates the tendency to experience negative emotions like anxiety and depression, along with overall emotional reactivity.

Within those five factors reside countless secondary traits. Extroversion can cover warmth or thrill seeking; neuroticism may include moodiness or a tendency to guilt. And as with the Myers-Briggs test, each factor contains its opposite: scoring low on extroversion means you lean toward introversion, scoring low on conscientiousness might mean you are sloppy or forgetful. The test itself usually takes the form of true/false statements—"I have frequent mood swings"—and the categories it deals people into do seem to connect to real-world outcomes. Studies have shown, for example, that individuals who score high on neuroticism tend to be less hopeful and more likely to exhibit depression. And chances are good that those who score low on openness to experience will be politically conservative. Research also suggests that those who score high on agreeableness may not be as successful at making money—though that may say more about today's business world than it does about human personality.

Of course, the drawback to boiling down personality to fit a spectrum of five factors is that, inevitably, something gets lost. Spirituality doesn't seem to be obviously accounted for in the scheme, and for the most part neither is sexuality. And that's only one of the flaws. As with other tests, there is also a subjective element to the scoring of this system, relying as it does in part on an analyst's interpretation. Likewise, there's the chance that subjects will answer less than truthfully.

But at an even more fundamental level, skeptical researchers have noted that the Five Factor Model presupposes that personality remains essentially unchanged over the years, and that doesn't jibe with the increasing amount of empirical evidence that suggests personality is much more fluid—that as we age, so does our outlook. It's a finding that should put lots of personality testers on notice. In the end, though, maybe we should be relieved that no single test seems equipped to accurately capture the complexity of the human psyche. Why would we think one could? After all, isn't it true that we behave inconsistently, even during a particular period? The person we are at the office is often not the person we are at home or out with friends. No, people will never be as straightforward as multiple-choice tests, and reliance on such tests—particularly in environments where an accurate reading counts most, like businesses and schools—can lead to unfortunate conclusions.

We are not prisoners of personality, whatever our type. The Harvard psychologist Brian Little always considered himself a deep introvert. Yet even though he's more comfortable in a library than on a podium, he has managed to build a quite public career as a lecturer and professor. Little calls this phenomenon—the ability to push through personality—Free Trait Theory. And what it comes down to is: while we all have certain fixed bits of personality—bits that might even be detectable by a test—we can also break out of character to pursue important goals, even if going against type causes discomfort. The key, he explains, is balancing three equal but very different identities. There's our mostly inborn personality; that's the biogenic identity. There are the expectations of our culture, family, and religion—the sociogenic identity. And there are our personal desires and sense of what matters—the ideogenic identity.

It might be easiest to live only according to our biogenic identity—that is, literally, our comfort zone. And behaving exactly as society demands minimizes conflict. But something will be lost if we fail to stretch ourselves beyond personality. "Am I just going to let things wash over me, or am I going to strike out and grow and change and challenge?" Little says. "Your response depends on what you want out of life."

And no test yet devised can adequately answer that for you.

Anger:
A Love Story

*We can't always get what we want—and how we learn to handle
that primal frustration, explains a noted Freudian psychologist,
is a crucial shaper of our personalities.*

BY HENRY KELLERMAN

NVIRONMENTAL INFLUENCES, we know, are tremendously important in how one's personality will eventually look. In fact, it is quite possible to trace or map specific influences that can make us or break us. When psychoanalysts examine the origin of one's development, they are actually looking for one word that would be the essence—the algorithm or fount—of what ultimately informs just about everything we think and do. To start such a search for this algorithm, this "source," shrinks ask us to remember. Getting to that early memory refers to one's history, to one's childhood. And when we do finally remember, is it possible that it's the same thing that we all remember? The answer is—yes. We will all remember the same thing.

The first thing to examine is what we're all required to do at the beginning of life, when we start to talk and to hear from our elders what is permissible to do and what the word "no" means. This period of our development marks the beginning of the concealment and repression of what psychotherapists need us to remember. In reasonably normal family life, parents are always concerned about helping their child learn the dos and don'ts of civilized living. Of course, it's the don'ts that really

count: "No, don't do that." And "No, you can't have that." And "No, don't touch the hot stove." For infants and young children, the "no" is always accompanied by an implicit "or else!" even when it is uttered in pearly tones. It's this implied threat and how the child feels about it that gives us our first clue about what the child conceals, even from herself.

The threat animates and encourages the child's compliance, even as she implicitly understands that the "or else" exists in the context of parental love and concern. The child's compliance to parental instructions, values, and style is the true engine that begins creating a strong sense of how to be. At the same time, however, something else is being created.

We need to understand that all children, whether they feel loved or not, are always in some form considering the idea of being abandoned. What makes the child feel completely helpless is even the remotest thought of being without parental protection. It is this apprehension and emotional vulnerability that gives the child an incentive and willingness to listen, to comply. For every person, though, even this vague sense of being controlled will generate anger or even rage, which is experienced, for the most part, unconsciously.

This is especially true for the young child, who is both completely at the mercy of external forces and not yet well able to determine what exactly is happening. In a

loving family it may appear to other eyes that the child gets, more or less, whatever he wants. But to this being who is still new to the ways of the world, it surely does not feel as though each and every wish is being met to the fullest measure. Through what is perceived inevitably as blocked wishes, the child experiences a pervasive sense of disempowerment. And that makes the child feel angry.

Especially in a young child, the most disempowered of beings, such anger becomes reinforced by every angry moment that arises from each thwarted wish. It has to be that way because, for all of us in moments of disempowerment, anger is always a re-empowerment. And all of us (including the youngest children) want to be empowered.

What happens with this anger? The child usually cannot show it, because the fantasy and fear of abandonment force it to be concealed. To be sure, children throw tantrums, but expressed anger rarely gets them what they want and eventually they learn to repress it. Typically, the child feels only that something is vaguely wrong, and that feeling quickly transforms into a sense of guilt and a sense of wanting to do good things to erase that bad feeling. From even a sort of incorrect sensation, then, a good result can occur: the child learns to do things the right way and become a good citizen. In the doing, she can't remember the apprehension and fear, and especially can't remember the repressed anger. The

> **We are all wish-soaked creatures. What this means is that we are all governed by what is known as the pleasure principle. We all want what we want when we want it.**

good news is that it doesn't matter much what the child remembers exactly. The same pattern will continue to occur throughout childhood, and then, believe it or not, throughout life.

That pattern ensures that while a person may feel dissatisfied with others, or feel anger toward them, for one reason or another he is not able to express that anger directly, or even, for that matter, be conscious of it. What needs to be remembered—where consciousness, as Freud says, can become curative—is a greater awareness of the fact that all of us are angry a lot of the time without really knowing it. Instead, we call it being upset, or being bored, or feeling moody or dissatisfied or stressed, or any number of other code words that describe what we are really feeling: angry.

And make no mistake about it, the anger is almost always caused by someone else who has prevented the angry person from getting something he wanted. Anger is always about the same thing: the thwarting or blocking of our wishes. And this wish-blocking constitutes the moment of birth in the development of a psychological or emotional symptom.

We are all wish-soaked creatures, all governed by what is commonly known as the pleasure principle: we all want what we want when we want it. When our wishes are met, we feel good and we feel empowered. When we don't, the opposite is true. The problem is that the world is constructed in such a way that al-

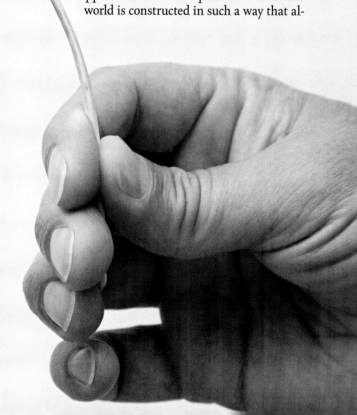

lows the existence of too many factors that can and will affect everything we do, and that means we are rarely able to control the situations in which we find ourselves. As when we are infants, we still can't seem to get what we want in the fullest measure, and our incomplete gratification leaves us either a little or a lot frustrated.

All that changes is the wish that is being blocked. When we are older, the thing that someone prevents us from obtaining is a job or a raise in salary. Or maybe it's sex, approval, recognition, or adoration. Or food or success. Whatever it is, and whenever it happens, your anger will most often not be able to be expressed toward the particular person who is frustrating you because that person is likely to be a parent, boss, partner, teacher, client, customer, important friend, and so forth. At times, your anger toward such a person may be socially almost impossible to express. So you instantly hide and repress the anger, burying it in your unconscious so that even you are oblivious to it.

The question then becomes: so what? We push down anger because, as in childhood, we don't want to or can't afford to let the person who is causing it to sense it. And we do that because we assume that should the person know about our anger, then he or she will … do what exactly? Well, we believe the other person will reject us, hate us, leave us—abandon us. That is what it all boils down to. Just as we were as children, we frequently remain afraid to direct our anger at those who hold our security in their hands. And this reaction is reinforced by the civilized standards of behavior by which we all live.

Basically, what's underneath is concealed anger, but what's on top is the rigmarole we concoct to help us get away from the fear of rejection or abandonment. These little strategies and tactics become the consistent behavioral patterns of ours that we use to help us massage situations so others will not see either the fear on top or the anger below.

It is these behavioral-personality patterns, built on the need to conceal underlying emotions, that get displayed in ways that determine our social persona. They become etched in us, and ultimately recognizable as our signature selves—for example, whether we are assertive to challenges confronting us or whether we are passive or timid in response to them. These patterns are the basic structure around which forms everything concerning the full nature of our personhood. No matter how we dress, how we feel, or what we do, these characteristics all hang from this basic skeleton of personality. It is our algorithm, the source of our thinking and behavior.

The fact that we are wish-soaked creatures is not in itself the problem. The real problem is that when the wish is blocked by someone else, we get mad. Making matters worse, most of us never make the distinction between major and minor wishes. We treat each wish as though it were major, such that not being able to find our wristwatch when we know it is somewhere in the bedroom is as frustrating and annoying as when we know we've actually lost it somewhere outside the home. Likewise, we fail to calibrate the difference between trouble and aggravation. Everything is treated as trouble. In this sense, we find ourselves wound up most of the time in wishes and troubles. If you know someone well enough, how she will respond to particular situations becomes predictable.

In the formation of anyone's personality, the issue of which dominates, impulse or control, is central to that person's particular style. Will she turn out to have an emotion-controlled style or an emotion-dyscontrolled style? (Emotion-controlled people sense that feelings can run amok, so control is perceived as an opportunity to feel safe and secure. Emotion-dyscontrolled people seek continuously stimulating situations because they feel more comfortable when their feelings are untrammeled.) It's clear that an impulse-dominated personality is, to some extent, immature, while a control-dominated personality is probably one that is better off. However, too much control in a personality can also be considered immature.

Individuals who are relatively adaptable display a comfortable fluctuation in the interplay between impulse and control. A normally functioning person will be able to develop an equilibrium, excluding those occasional shifts in which, depending on changing circumstance, one or the other side needs to dominate. A decent balance is needed to ensure that the personality will be one that can withstand pressure, that can delay gratification, and that can be resilient and able to withstand the daily pressures of life.

One common impulsive form of response to painful feelings is acting out, rather than talking them out (which serves the purpose of making us more conscious of them). Acting out is doing rather than knowing. And when people continuously do something that is undermining to the self as well as to others—for instance, lying, stealing, being sexually promiscuous—it's likely that their psyche is keeping them in a steady state of external stimulation to prevent a state of insight. The bottom line is that the psyche is insisting that the thwarted wish be gratified in a neurotic or perverse way. Essentially, it has translated the wish into a symptom. As Freud told us, we all love our symptoms because they are our wishes fully realized.

Interestingly, in psychotherapy the only thing that can get cured is a symptom—perhaps a phobia or an obsession or an intrusive thought. All else in psychotherapy is an attempt to assist the patient in the ability to struggle better with life—with greater effectiveness and success. Life does not get cured; only symptoms can be cured. Struggling better is the goal.

Henry Kellerman, a psychologist and psychoanalyst in New York City, is the author of many books, including Personality: How It Forms *(2012), from which this essay is adapted.*

The Myth of the Alpha Dog

Turns out what holds in the animal kingdom may not work for humans. The leaders of our packs are made, not born.

BY MICHAEL Q. BULLERDICK

THAT DIFFICULT, DEMANDING boss who still haunts your nightmares? Blame him on the wolves. Turns out man's earliest concept of leadership originated with our wild canine friends. Way back when, 135,000 years ago or so, humans and canines developed along parallel and codependent tracks. While we were busy transforming wolves into domesticated dogs, they were busy teaching us a thing or two about structuring an efficient society. Observing wolves, early humans quickly figured out they stood a better chance of survival if they formed close-knit, hierarchical packs that were managed by the strongest, smartest, and most aggressive among them—the alpha dogs.

After so many intervening generations of survival-based psychosocial reinforcement, a command-and-control model of leadership remains an entrenched part of humankind's collective unconscious. Which is why, for the better part of the 20th century, the study of leadership dynamics was principally concerned with identifying common alpha traits. Good leaders, psychiatrists asserted again and again, were born of the best stock, with certain heritable characteristics that pegged them as ideally suited to the task of directing others: extrover-

sion, decisiveness, charisma, and, yes, even a certain level of aggression, to name a few. The prevailing wisdom was that you either had what it took to succeed or you didn't—end of story. For nearly 50 years the U.S. military and many Fortune 500 companies recruited and groomed potential officers and executives based on entrance and performance evaluations designed to reveal desirable traits and weed out disqualifying flaws.

But though it was perfectly plausible as a research starting point, trait theory fell flat over time. As researchers learned more about brain function and the more complex origins of personality, they saw that their understanding of leadership was insufficient. For one thing, trait theory failed to locate the origins of leadership traits in the brain. It also failed to explain whether these traits could be acquired or enhanced. Over the last 20 years or so, however, newer technologies, including brain imaging, have allowed the science to go beyond trait theory, revealing some fundamental principles about brain function that may one day send the concept of alpha dogs to the pound.

Trait theory took its first debilitating hit when research revealed that the brain does not settle into a relatively fixed state as we grow into adulthood. Rather, it remains malleable throughout our lifetime. Neuroplasticity, as the scientists call it, covers how our thoughts and actions and various external stimuli alter both our

MOHANDAS GANDHI *Holding no official title and commanding no armies, Gandhi led India toward independence. "A man is but the product of his thoughts," he said. "What he thinks he becomes."*

ABRAHAM LINCOLN *"You cannot build character and courage by taking away a man's initiative and independence," said Lincoln, whose ability to motivate and empower people grew over the course of his career.*

ELEANOR ROOSEVELT *Not viewing herself as a leader, Roosevelt attributed success to empathy and learning from failure. "The influence you exert," she said, "is through your own life and what you've become yourself."*

MARTIN LUTHER KING JR. *"The supreme task [of a leader] is to organize and unite people," said King, who believed that wisdom came from even the most humble people and no one was to be underestimated.*

brain's structure and its function. It works like this: When we commit to learning something new—say, playing a musical instrument—neurons in affected regions of the brain seek out other neurons, interconnecting to create waypoints of stored memory. When a critical mass of neurons unite, a new network is formed. The network will continue to become stronger and faster the more frequently it is accessed. This explains not only how we learn to play an instrument at all but also why we are able to master it over time.

Neuroplasticity demonstrated that the age-old question about whether leaders are born or made was largely irrelevant. If the brain can bypass unwanted, insufficient, or damaged neural pathways to build new and improved ones, then the possibilities for self-improvement could be limitless. The follow-up punch to alpha-trait theory was the realization that positive learning takes place only when learners feel safe, mentally in tune with their leaders, and truly engaged in a task. Once again it was brain-scanning technology, like functional magnetic resonance imaging (fMRI), that was responsible for updating the model. Measuring brain activity, researchers discovered that critical thinking and learning are inhibited by fear. Fear, they found, channels resources away from higher brain regions and toward a lower region known as the amygdala, a part of the limbic system, which processes emotions. The amygdala goes into overdrive in response to threats, activating our primal "fight or flight" reflex that floods the body with adrenaline and cortisol. As a result, brain functions that facilitate learning—such as cognition, rational thought, and logic—take a back seat and are effectively suspended. The same holds true when we experience milder forms of fear like anxiety and stress. It's why office tyrants can succeed in motivating subordinates to perform rote tasks, such as reporting for work on time, but routinely fail to inspire creativity.

Experts will say that leadership, boiled down to its essence, is really the ability to communicate goals and influence people to act in ways that make achieving those goals possible—to motivate in order to effect change. That's sounds straightforward enough, but it's easier said than done. "Try to change another person's behavior, even with the best of intentions, and he or she will experience discomfort. It takes a strong will to push past such [fear-based] mental activity," note researchers David Rock and Jeffrey Schwartz, titans in the study of leadership dynamics. "The brain sends out powerful messages that something is wrong. Change itself thus amplifies stress and discomfort."

So if fear and intimidation won't cut it—certainly not in today's information-age industries—what will? It turns out the best leadership strategy is a collaborative approach. Psychologist Daniel Goleman, a pioneer in fMRI research, talks about "resonant leaders," those men and women who demonstrate an emotional responsiveness toward their subordinates as they engage them in the creative and decision-making processes. When they impart information—by speech, body language, or energy output—resonant leaders continually monitor their own actions even as they interpret the emotions of their followers. Spotting subtle signs of rejection or disengagement, they are able to make on-the-fly adjustments to keep the creative processes firing. The term for it, emotional intelligence, was popularized by Goleman, but the concept goes all the way back to the work of Charles Darwin.

Abraham Lincoln, who had the toughest leadership task of any American president, demonstrated "an extraordinary amount of emotional intelligence," wrote historian Doris Kearns Goodwin in a 2009 essay about the 16th president's management style in the *Harvard Business Review*. Although he was largely self-educated and certainly received no formal leadership training, Lincoln transformed himself from an ineffectual congressman from Illinois to the shrewd strategist who navigated the country through its worst political crisis. Credit his success, Goodwin says, to that emotional resonance. As her book *Team of Rivals* chronicles, Lincoln had the confidence to appoint former enemies to cabinet posts and encouraged lively debate with dissenters who never had to fear rebuke or retaliation. More important still, he listened well and empathetically as he opened the White House to the public and waded among the crowds to take what he called "public-opinion baths." Lincoln claimed he rarely gave direct orders; instead, he preferred "offering advice as suggestions" and "telling little stories,"

> "Be the change you wish to see in the world."
> —*Mohandas Gandhi*

often incorporating humor, to turn others toward his way of thinking. "If you would win a man to your cause," he said, "first convince him that you are his friend."

If neurologists had been around in mid-19th-century America, they would have written papers that linked the success of Honest Abe's invitational approach to a brain phenomenon known as neural mirroring. The concept explains what occurs when neurons within major brain centers of different people fall into alignment. Goleman's 2008 study, also published in the *Harvard Business Review*, concluded that resonant leaders "help activate openness to new ideas [neuroplasticity] and a more social orientation to others [neural mirroring]." Earlier groundbreaking work published in *Current Biology* revealed that during this optimal state, participants mentally mimic their leader's actions, emotions, *and* intentions. It's what we think of more commonly as trust, and it occurs when the brain boosts production of oxytocin, a hormone that facilitates bonding—between mother and child, between mutually attracted individuals, between long-term friends and tribesmen.

Clearly, neural mirroring can increase the odds that shared goals will be met. More significantly, though, initiating the process cultivates the growth of receptive pathways in the brains of subordinates. That's neuroplasticity—*learning*—at work. One especially intriguing aspect of neural mirroring is that it can even occur remotely, for instance through videoconferencing. That's no small thing in today's global workplace. What matters, say researchers Boris Groysberg and Michael Slind, is "mental or emotional proximity." Adept leaders, they note, "step down from their corporate perches and then step up to the challenge of communicating personally and transparently."

At his standing-room-only seminars, Harvard psychologist Srinivasan S. Pillay teaches would-be leaders just how to step up. Like the rare magician who reveals the mechanics behind his deception, Pillay lectures his audience on the principles of neural mirroring, then demonstrates its immediate effectiveness. In short, he uses neural mirroring to teach it. "The brains of these listeners are suddenly faced with an understanding of themselves," writes Pillay in his book *Your Brain and Business: The Neuroscience of Great Leaders*. "And I can sense that there is a readiness to change." As minds are transformed, a new breed of boss is made.

Pillay has built a successful business around preaching that if we can't all magically morph into latter-day Lincolns, we can learn to be more collaborative, emotionally resonant managers. And that means while some top dogs are undoubtedly born, others can surely be molded. Sometimes all it takes to lead is the willingness and fearlessness to follow a new approach.

Disorders

The effort to identify and treat a range of troubling personality conditions is one of the vital frontiers of mind science. Do we overpathologize human behaviors? There are those who would say the opposite is true. (And, no, we are not talking about you.)

{
PAIN, RAGE, AND BLAME
IS THERE AN ADDICTIVE TYPE?
OUR CULTURE OF NARCISSISM
NEW CLUES TO OCD
A VOTE FOR LABELING
}

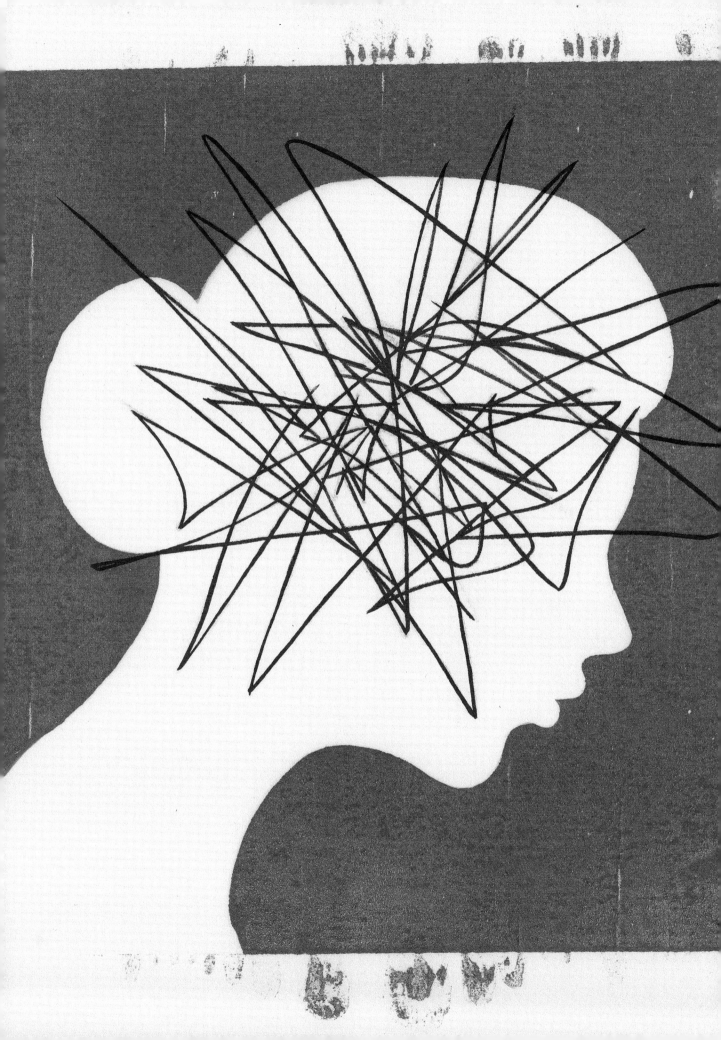

Pain, Rage, and Blame

Personality disorders are among a psychologist's toughest cases. Other patients know they have a problem; these patients insist everybody else does.

BY JEFFREY KLUGER

NO ONE GOES CRAZY ACCORDING to schedule, but it's long been a sure thing that in May 2013, a lot of people would. Fortunately, just as many already-crazy people might get instantly well. In either case, crazy is probably too strong a word. But a lot of other words—volatile, vain, miserable, terrified, inflexible, impulsive, even delusional—work just fine. Those and other unhappy adjectives accurately describe the millions of people suffering from the cluster of 10 conditions known as personality disorders. These include some familiar psychic ills (narcissism, paranoia, histrionic behavior) and less familiar ones (borderline and schizotypal personalities). None, however, are any fun at all, either for sufferers or the unlucky people in their orbit, and all can be a devil to diagnose—presenting as a messy hash of overlapping and comorbid symptoms.

It's in May 2013 that the fifth edition of the Diagnostic and Statistical Manual of Mental Disorders (DSM-5) was set to be released, and in the years preceding its publication the authors labored long to try to streamline the section on personality disorders, putting all the conditions on a sort of sliding scale that would allow for patients to be diagnosed more by severity of symptoms and less by a discrete label. Untold numbers of people who thought themselves reasonably healthy might have been snared in the new diagnostic net—and untold numbers who were being treated for conditions might have been tossed out. In the end, that danger—not to mention the maddening conundrum of determining just what personality disorders are—forced the authors to punt; keeping all the same diagnostic categories, they merely tweaked the definitions and promised to study it all further.

Still, praise them for trying. And pity the psychologists and psychiatrists who step up to treat folks with personality disorders at all. For those patients are the ones who show up to therapy in the first place only because their family or friends or co-workers have finally had enough of them. Fact is, people with personality disorders really, truly believe that they're just fine, and that it's everyone else who is nuts. They're not paranoid; people actually are after them. They're not histrionic; they're just expressing themselves in the only way the rest of us thick-headed lugs can understand. They're not antisocial; they're just being themselves. They're not narcissistic; they simply are that great. "They come in under duress," says Lawrence Josephs, a psychology professor at Adelphi University in Garden City, N.Y. "But what they really want is to have everything on their own terms."

Although it's surely no comfort to Josephs, he is not

{Problem Profiles}

The 10 currently recognized personality disorders often manifest as extreme versions of normal behavior.
(An 11th catch-all category exists for cases that don't fit the other diagnoses.)

Avoidant
A sense of inadequacy; extreme sensitivity to being seen as flawed; fear of social interaction.

Paranoid
Suspicion and mistrust of others.

Schizoid
Not schizophrenia (a full-blown psychosis); social isolation and limited range of emotional expression.

Schizotypal
Not schizophrenia either, but getting dangerously close; similar to schizoid, but with some delusions.

Antisocial
Disregard of the law and the rights of others; a tendency toward deceit and manipulativeness.

Borderline
Impulsiveness and volatility; unstable relationships; often self-destructive or suicidal tendencies.

Histrionic
Extreme emotionality and attention-seeking; may include excessive sexual seductiveness.

Narcissistic
Grandiosity and self-absorption; an insatiable need for recognition; a lack of interest in others.

Dependent
Clinginess and submissiveness; difficulty taking responsibility for decisions or disagreeing with others.

Obsessive-Compulsive
Not to be confused with OCD (an anxiety disorder); defined by rigidity and overadherence to rules.

alone in having such trouble with certain patients. A clinical personality disorder (PD) is one of psychology's toughest nuts to crack. Talk therapy often doesn't touch the problem; drug therapy may do just a little better. And disturbingly, as families increasingly fragment and societal pressures grow, experts say they are seeing more and more cases. As much as 9% of the population is thought to suffer from some kind of personality disorder, and as many as 20% of all mental-health hospitalizations may be a result of such conditions. "The more severe cases are increasing," says Josephs, "especially among people who grew up in homes with divorce or drug and alcohol problems."

There are a lot of reasons that personality disorders are so resistant to treatment, but the most important may be the way they insinuate themselves into the mind of the sufferer. Other mental conditions, such as anxiety disorders and depression, can be imagined as a pathological rind wrapped around an intact core. Peel the skin away with talk therapy or melt it away with drugs, and the problem often abates. Personality disorders, by contrast, are marbleized throughout one's entire temperament. At the same time, their impact extends well beyond the sufferer. "The social costs of personality disorders are huge," says John Gunderson, director of the Personality Disorders Service at McLean Hospital in Belmont, Mass. "These people are involved in so many of society's ills—divorce, child abuse, violence."

While solutions are elusive, the pathological arc of PDs is predictable. Symptoms tend to show up after age 18, striking men and women equally, though gender may influence which of the 10 disorders a person develops. The conditions are grouped into three subcategories, and of these, the ones in the so-called dramatic cluster—borderline, antisocial, narcissistic, and histrionic disorders—are the best known and the most explosive. And of that quartet, borderline is the most troubling and least tractable.

Borderline personality disorder got its precarious-sounding name because sufferers were once thought to live at the frontier at which an ordinary disorder gives way to a full-blown psychosis. And while that turned out not to be really the case, it's easy to see how the mistake was made. In the context of interpersonal relationships, borderline patients are in a constant whipsaw of emotion—deeply, often unrealistically adoring of the people they love, while just as primally fearing abandonment. It leads to both devotion and explosions, tenderness and a sort of emotional terrorism. When borderline patients want to hug, they claw instead, and that can make them not only impossible to live with but impossible to treat. Says Kenneth Duckworth, medical director of the National Alliance on Mental Illness: "It used to be that if you hated the patient, you would bandy this term about: 'Oh, you're just a borderline.' As a medical diagnosis, it was a wastebasket of hostility."

That's not so anymore, thanks to the pioneering work of University of Washington psychologist Marsha Linehan, who has unforgettably described borderline patients as "the psychological equivalent of third-degree-burn patients. They simply have, so to speak, no skin. Even the slightest touch can cause emotional suffering."

To ease the torment, she created—and has widely taught—a therapeutic technique known as dialectical behavior therapy (DBT). "Dialectical" implies the tension that lies in the contradictory truths that both involved parties need to embrace. Doctors must accept the patients' rage, hostility, and generally intolerable behavior—those same characteristics that led to an earlier generation's professional hostility—while committing to the work of helping them change. Patients must end their tendency for black-and-white thinking, while accepting that some behaviors (principally their own) are simply unacceptable. The therapy takes a lot of work, but it bears fruit. According to one study, 88% of patients who receive a borderline diagnosis no longer meet the criteria for the disorder a decade after starting DBT.

Less volatile than the dramatic disorders are those in the so-called anxious cluster: the straightforwardly named dependent personality, the socially withdrawn avoidant personality, and the rigid and rule-bound obsessive-compulsive personality disorder (OCPD). This last one is among the most poorly named of all conditions, as it is obviously and endlessly confused with common obsessive-compulsive disorder (OCD), an anxiety condition. Better to think of OCPD as perfectionism, even if that doesn't quite capture it, because sufferers expect nothing less than precise and flawless work from themselves and are often unable to complete projects because nothing is ever good enough. How do you finish a 40-page business memo if you can never get past polishing page one?

Worse is when the same people turn their exacting standards outward. OCPD sufferers are the ones yelling the loudest at the airport counter agent or fuming the hottest when they're stuck in traffic, because flights should not be canceled and a freeway should jolly well move like one. In the world the rest of us live in, however, what should be is rarely what is, and only when patients learn that lesson will they get well.

The third group—called the odd cluster—covers the paranoid, schizotypal, and schizoid personalities. Paranoid sounds just like what it is. Schizotypals and schizoids both have problems forming relationships and interpreting social cues. Schizotypals may also suffer delusions. "Schizoids are lone wolves," says Norman Clemens, emeritus professor of psychiatry at Case Western Reserve University in Cleveland. "Schizotypals skate along the edge of real schizophrenia."

Before scientists can settle on how to treat these conditions, they'll have to figure out just what is behind them. Few researchers doubt that when disorders are so woven into temperament, some of what causes them must be written into genes. A Norwegian study published in 2000 examined identical and fraternal twins and found that matched pairs—with their matched genetic blueprints—were more likely than unmatched pairs to share personality disorders. The borderline personality had an estimated 69% level of heritability, for example.

But genes aren't everything; in fact, some investigators doubt they play much of a role at all. Twin studies can be inherently unreliable, not least because fraternal twins have a tendency to try to accentuate their differences while identicals tend to highlight their similarities. Those response biases can fatally contaminate the answers to questionnaires that often lie at the heart of research like the Norwegian study. What's more, research like this does not isolate any actual genes linked to a trait—it merely uncovers statistically significant patterns that suggest those genes exist. Even when genes are pinpointed, however, as they already have been in the case of some physical illnesses, they rarely play more than a 10% role in determining whether the person carrying them actually develops the disease.

More powerful than heredity, at least in personality disorders, is environment. Therapists who work with narcissists often uncover childhood abuse or some other trauma leading to low self-esteem or even self-loathing—just the kind of emotional hole that pathological grandiosity would be designed to fill. This is known, straightforwardly enough, as the mask model. Then there are psychologists who believe just the opposite: that narcissism is the result of a childhood spent being overpraised and adored, making it hard to adjust to a world in which plaudits do not come so easily.

Borderline personality disorder (BPD) affects more women than men, and some research has shown that up to 70% of borderline women were sexually or physically abused at some point in their lives. Sometimes something as simple but destructive as being forbidden as a child to show emotion—*if you must cry, go to your room and do it there*—can get the dysregulation that leads to BPD going. Poorly handled bipolar disorder or learning disabilities can also evolve into personality disorders. Even full-blown schizophrenia (a psychosis, not a personality disorder) can be exacerbated or ameliorated by upbringing and family stability, and this is surely true of its milder cousins, schizoid and schizotypal, as well. Still, the connections are not always clear, and that can make diagnosis—to say nothing of therapy—harder still.

Truthfully, the trio of clusters themselves are, at best, imperfect definers of the conditions they are meant to help diagnose, and that's the problem the DSM-5 authors were trying to rectify. Avoidant personality, for example,

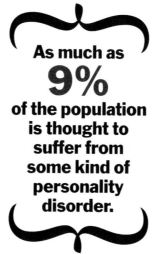

As much as 9% of the population is thought to suffer from some kind of personality disorder.

shares a lot of traits with social phobia, an anxiety condition, and yet they're different too. Dependent personality is, in some ways, halfway to borderline personality—but it lacks the volatility. And the fact that OCD and OCPD are so different doesn't mean perfectionism and a sense that things have to be just so aren't applicable to both. It's simply that the reactions—anxiety in one case, rage in the other—are different.

But whatever the specific roots of these conditions, once the environmental and genetic dice are cast, that is often that for the disordered personality—at least as long as those afflicted continue to resist acknowledging their problem. Anxiety disorders such as phobias are generally labeled ego-dystonic illnesses: the sufferer knows that a problem exists and wants—sometimes deeply—to do something about it. Personality disorders are ego-syntonic: the individuals believe their behavior, however destructive, is appropriate. Unclear thinking like that makes a patient hard to heal. But there is hope, and some of it is being created in the pharmaceutical lab.

Researchers are finding that antipsychotics can help alleviate paranoid, schizoid, and schizotypal symptoms. A variety of drugs—including mood stabilizers, anticonvulsants, and SSRIs—may help control the impulsive element of the dramatic disorders. And while antidepressant and antianxiety medications do little to rejigger something as fundamental as personality, doctors find that if these drugs relieve the stress that comes with living so chaotic a life, some motivated patients may be willing to take on the harder work of talk therapy.

For those who do, the options are growing. As with borderline patients who accept DBT, cognitive and behavioral therapy in particular can teach coping skills that first help people gain a more honest view of their situation, then fix what's not working. A study conducted by Gunderson and his colleagues at Harvard University and elsewhere found that after two years, borderline, avoidant, obsessive-compulsive, and schizotypal patients showed a 40% improvement. "That's big news," says Gunderson. "Nobody thought we'd get better than 15%."

Forty percent improvement, however, still leaves 60% in distress, so researchers continue to work to tip that painful balance the other way. Until they do, it will mostly be up to patients to deny the lie that their disorder tells—that nothing is wrong with them—and make the therapeutic commitment that is necessary to get well. "Nobody totally changes," says Josephs. "But anyone can become more flexible and resilient. Anyone can make progress." That is already a better prognosis than most of these patients have traditionally had.

Redefining Mental Illness
New guidelines will change how we assess what ails the mind.

IN THE WORLD OF MENTAL health, the Diagnostic and Statistical Manual of Mental Disorders is more or less the bible. Doctors use the DSM's definitions to diagnose depression, stuttering, fetishism, schizophrenia, and more than 300 other conditions. Insurance companies use it to justify reimbursements; without a DSM code, mental-health patients usually don't get a dime. And the manual carries enormous cultural heft: when it stopped listing homosexuality as a mental disorder—after a 1974 psychiatrists' debate in which being gay was deemed sane by a vote of 5,854 to 3,810—gay rights received a crucial boost.

So naturally, on December 1, 2012, when the American Psychiatric Association's board of trustees approved a fifth edition of the DSM, which took 13 years and 1,500 mental-health experts to complete, it rocked the medical world. But the APA's approach—treating mental disorders less as discrete illnesses, like leukemia, and more as problems on a continuum, like hypertension—also clarified its intellectual approach and expanded its reach. Here's how the new classifications will affect patients.
—By John Cloud

Sources: American Psychiatric Association; Mayo Clinic; University of Washington

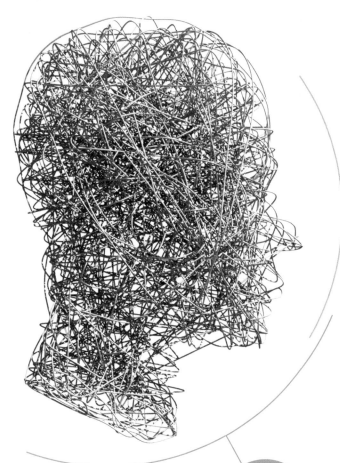

GOOD NEWS FOR ...

HOARDERS
For the first time, hoarding disorder will be included as a diagnosis, so those who can't get rid of ephemera will be able to seek reimbursement for therapy.

BINGE EATERS
The DSM has listed binge-eating disorder in an appendix for more than a decade. From now on, it's an official diagnosis.

THE BEREAVED
The previous DSM said those in mourning don't necessarily qualify for depression therapy or medication. DSM-5 eliminates that exclusion.

SKIN PICKERS
Psychiatrists have long debated whether excoriation, or skin picking, should be considered a mental illness; in DSM-5 the debate has been resolved, and the answer is yes.

MIXED NEWS FOR ...

AUTISTICS
Although it reclassifies autistic disorder as autism spectrum disorder, which includes Asperger's, the new DSM-5 definition doesn't do much at all to improve doctors' understanding of autism.

BAD NEWS FOR ...

DIVORCED PARENTS WITH DIFFICULT KIDS
Years were spent debating a proposed diagnosis of parental-alienation syndrome, the difficulty kids feel after parents divorce. Despite much outrage, DSM-5 did not include it.

SEX ADDICTS
DSM researchers rejected the idea that hypersexual behavior is a mental disorder, making it tough for those with extreme sexual urges to seek treatment.

Who Gets Addicted?

People with bad habits come in all varieties. But finding their particular strengths is the path to treatment.

BY DAVID BJERKLIE

FOR ALL THE HAVOC IT WREAKS, you'd think addiction would have a definition that was just a tad more dramatic than this: "continued use of mood-altering substances despite adverse consequences." Then again, maybe it isn't possible to describe with more menace something so diffuse. Researchers, armed with increasingly sophisticated technology, have made great strides in understanding the details of what goes wrong in the brain of an addict. But a good portion of the story lies beyond the neurotransmitting chemicals and regional activation of brain cells and genetic vulnerabilities. All that biology is invariably entangled within a complex web of social and psychological factors.

The question of whether there is a classic addictive personality is a longstanding one, and it's more than reasonable to believe there might be. After all, don't addicts tend to be more impulsive and less able to delay gratification than those less inclined to succumb to dependency? And don't similar insecurities, needs, and fears lie at the heart of many addictions? On the other hand, don't addicts represent every type of personality imaginable?

Aren't they shy and bold, boorish and sensitive, funny, dour, angry, passive, warm, chilly, stingy, generous, staid, stubborn, and submissive?

Unhelpfully, the answer to each of those questions is yes. But while there may not be one profile that fits all, what researchers have found is that there are traits associated with a higher risk of addiction—and traits that appear to protect individuals from addiction too. Even 30 years ago, the National Academy of Sciences concluded that while there was no single set of psychological characteristics that pegged all addicts, there were several significant ones that showed up again and again, including impulsivity, nonconformity, and social alienation. More recent studies have found that individuals with personality disorders appear to have a higher prevalence of alcoholism and other addictions. But the personality-addiction connection is a two-way street: while certain personalities may increase the risk of addiction, chronic substance abuse can also alter personality or exacerbate potential psychological problems.

Just for good measure, personality can also influence a person's prospects for recovery. "Although there is no evidence for a single addictive personality," wrote addiction specialist Daniel Angres in a 2010 article in the psychiatry journal *Focus,* "there are a multitude of ways

ADDICTED TO...

ALCOHOL: Almost 19 million Americans are dependent on alcohol; 12,000 try it for the first time each day. Recovery rates are low, relapse is the norm, and only a fraction of those who need help seek it.

DRUGS: About 3.6 million Americans are dependent on such drugs as marijuana, cocaine, and pain relievers; fewer than 20% are undergoing treatment. More than half of first-time users are female teens.

CAFFEINE: The world's most widely used mood-altering drug—ingested by 80% to 90% of Americans in coffee, tea, and sodas—can cause physical dependency and withdrawal.

GAMBLING: Two million American adults are thought to wager heedlessly. An additional 4 million to 8 million are considered problem, if not pathological, gamblers.

personality functioning affects addiction and treatment outcomes." It follows that isolating relevant factors—particular emotional strengths, say, or a capacity to change—is a first crucial component to developing an effective treatment plan.

When it comes to addiction, though, the operating assumption has to be: expect variety. A 2007 study by the National Institute on Alcohol Abuse and Alcoholism concluded, in the words of researcher Howard Moss, "Our findings should help dispel the popular notion of the 'typical alcoholic.'" Moss and colleagues identified five major subtypes of problem drinkers. Some were characterized by higher than average rates of depression, anxiety, antisocial personality, or bipolar disorders, and family histories for the dependency. Overall, though, the majority of study participants had neither mental disorders nor a history of alcoholism.

The neurology of addiction almost guarantees that the population of those afflicted will be diverse. Addiction insinuates itself by hijacking the very brain functions that are critical to human survival, the reward pathways that have evolved to reinforce the behaviors—eating, drinking, having sex—that ensure we make it through another day, and another generation. Neurotransmitters such as dopamine are released when we engage in these behaviors, to produce feelings of pleasure and satisfaction that make us want to do them again and again and again.

Drugs and alcohol, unfortunately, also work this neurological road. And we're not talking about one simple reward pathway; multiple pathways, multiple neurotransmitters, and multiple regions of the brain are involved. So it's no wonder we're primed to get hooked. When we are exposed to addictive substances, our memory systems, reward circuits, decision-making skills, and conditioning all conspire to create rapidly and strongly reinforced patterns of craving. So, yes, addiction is clearly pathological; the great equalizer is that its underlying mechanisms are common to us all.

FOOD: Although food addiction is not synonymous with obesity, it has been implicated in binge eating. Stress, anxiety, and depressed mood can trigger bouts of compulsive eating behavior.

SHOPPING: One in 20 Americans may be a compulsive shopper. Behavioral addictions like compulsive shopping and gambling have symptoms similar to substance abuses.

TOBACCO: Upwards of 70 million Americans use tobacco products; 23.4% of men and 18.5% of women smoke cigarettes. The age group with the highest rate of use is young adults (ages 18 to 25), at 44.3%.

SEX: Sixteen million Americans (a third of them women) are estimated to suffer from compulsive sexual behavior, including a pornography fixation; about 60% of them were abused in childhood.

But Enough About You . . .

A little self-absorption is a good thing. Too much is a turn-off. Here's how to tell if you're just too into you.

BY JEFFREY KLUGER

THE MOMENT YOU WERE BORN, you fell in love—with you. You were really all you knew, after all: a sensory ball at the center of the universe, tended to by out-of-focus giants who had the power to keep you warm and cozy, to replace a wet diaper with a dry one, an empty belly with a full one. Those primal needs came first, before anything else—before anyone else—and that was just fine with you. You were born a perfect narcissist.

It's not for nothing that you came into the world in a state of utter self-absorption. You wouldn't have survived infancy if every time you needed something you gave even a moment's thought to the competing needs of the grown-ups who had to provide it for you. It's not for nothing that some of that me-first behavior remains throughout our lives. The world is a fang-and-claw place still—with its fixed amount of money, mates, and other resources—and often the only way to get yours is to make sure others don't get theirs.

Narcissism, like it or not, is adaptive, and the same goes for the egotism that is such a big part of it. You can't exactly convince potential employers, friends, and romantic partners that you're smarter, funnier, or prettier than the next person unless, at some level, you believe it too. But there's narcissism and then there's Narcissism— full, florid, capital-N narcissism. That's a whole different beast—and it's on the rise. In 2008, a team of researchers published a study in the *Journal of Personality* that looked at narcissism rates among college students over a 27-year period that ended in 2006. The subjects were all administered the Narcissistic Personality Inventory (NPI), a 40-item questionnaire that requires choosing between such essentially opposite statements as "I insist upon getting the respect that is due me" and "I usually get the respect that I deserve"; "If I ruled the world it would be a better place" and "The thought of ruling the world frightens the hell out of me"; "I am more capable than other people" and "There is a lot that I can learn from other people."

The results were striking: in little more than a generation there was a 30% increase in overall NPI scores, with more than two-thirds of subjects from 1986 to 2006 scoring higher than the mean of the period between 1979 and 1985. "The exact same test has been given every year," says psychology professor Brad Bushman of Ohio State University, one of the coauthors of the study, "and the narcissism rates are increasing all the time."

But we hardly need peer-reviewed research to inform us that our culture has taken a sharp turn into extreme self-love. There are the political figures and their serial

sex scandals—men moved more by a sense of entitlement than by any real impulse to serve the people they presume to lead. There are the branded celebrities with their product lines and nonstop Twitter feeds, broadcasting such ephemera as when they went to bed last night and what they had for lunch today. ("Just killed back to back spin classes," tweeted Lady Gaga. "Eating a salad and dreaming of a cheeseburger.") There's Donald Trump with his four-story name on everything he builds and his vanity flirtation with the presidency. There's Kim Kardashian with her sex tape and her reality-show empire and her self-named fragrances. "Acting and singing aren't the only ways to be talented," she told *The Guardian.* "It's a skill to get people to really like you for *you,* instead of a character written for you by somebody else."

But who are we kidding? There's the rest of us too—500 million posting our own irrelevancies on Twitter, a billion doing the same on Facebook. We have our own blogs and websites, our own vanity presses and vanity albums. And should we ever need a tune-up to our self-esteem, a tsunami of books beckons—68,000 of them on Amazon.com in paperback alone, all with names like *I Like Me!*; *The Best Part of Me*; *Happy to Be Me!*; and *The Loveables in the Kingdom of Self-Esteem.*

Somewhere between the poles of humility and grandiosity, it seems, we've gotten lost. The healthy self-centeredness of babyhood has grown into the stubborn self-centeredness of childhood and, ultimately, a sort of permanent self-centeredness of adulthood. One of our more adaptive traits is rapidly becoming one of our more destructive ones. But if our culture-wide narcissism is hard to deny, the reason for it is, at the moment, even harder to parse.

The pandemic may be a comparatively recent phenomenon, but the disorder—at least in the clinical sense—has been around for a while. Since the late 1960s, when the term "narcissism" replaced "megalomania" in some academic texts, it has been listed as one of the 10 clinically recognized personality disorders—an unhappy collection that includes histrionic, dependent, antisocial, avoidant, and borderline personalities. Full-blown clinical narcissism is defined by grandiosity, arrogance, lack of empathy, exploitativeness, and an overweening sense of entitlement—the belief that greatness and recognition are owed more than earned.

You'd think that a person exhibiting such a nasty mashup of disagreeable traits would learn pretty quickly that almost everyone else wants no part of it, but the narcissistic temperament has powerful ways of protecting itself. For starters, narcissists are often blissfully blind to how others see them, registering only the most favorable social cues while ignoring or misinterpreting the others. Ever sit through a dinner party during which one person holds forth insufferably from soup through dessert, seemingly unaware of the glazed eyes and surreptitious watch-glancing all around? It seems impossible to miss social cues like that—and yet there's a reason the NPI questionnaire includes such statements as "Everybody likes to hear my stories," "I like to be the center of attention," and "I will usually show off if I get the chance."

What's more, in some cases narcissists really do exude the charm and seductiveness they think they do. It's what often helps them succeed professionally, socially, and especially romantically—at least at first. "You meet someone extroverted and confident and you find it very exciting," says psychologist Keith Campbell of the University of Georgia. "But the relationship with the narcissist won't translate into intimacy and caring."

Instead, Campbell says, it will inevitably curdle into something unilateral, the narcissistic partner demand-

> **"We've shot more episodes than *I Love Lucy*! We've been on the air longer than *The Andy Griffith Show*! I mean, these are iconic shows."**
> —*Kim Kardashian, defending her fame in* The Guardian

ing attention and devotion while offering little in return. And that's true in nonromantic situations too. The narcissistic boss or co-worker might dazzle you early on with so much energy and charisma, only to have exactly the opposite effect on you later. "I think of it as the chocolate-cake model," says Campbell. "It's a rush in the short term, but after you eat it you want to kill yourself."

Happily, clinical narcissism is relatively rare—affecting only about 3% of the population. But just as there is a sliding scale down which disabling depression gives way to chronic melancholy, and agoraphobia becomes merely social anxiety, so too there are graduated degrees of narcissism—and this is where the disorder becomes democratized. Not everyone can be a narcissist on the order of a John Edwards or a Tiger Woods, but plenty of us can exhibit too much self-love for our own good, not to mention the good of the people around us.

The roots of narcissism—even the less extreme kind—are difficult to pin down, though genetics almost assuredly plays a role, as it does with so many other psychic ills. A 2000 study conducted at the University of Oslo compared the incidence of all the personality disorders in identical twins (who have matching DNA) with the incidence in fraternal ones (who have no more DNA in

common than ordinary siblings do). The researchers concluded that 58% of all cases of personality disorders could be attributed to genes. Paranoia, at 30%, was the least heritable; obsessive-compulsive, at 78%, was the most. Narcissism, at 77%, was a close second. Those are compelling numbers, certainly—but not, apparently, entirely convincing ones.

Social psychologist Jerome Kagan of Harvard University argues that twin studies are inherently unreliable because, he has found, fraternal twins typically highlight the differences between them while identical twins tend to highlight the similarities, which would dramatically skew the answers to the questionnaires and interviews that form the basis of these studies. What's more, inferring that genes are responsible for a disorder based even on honest and accurate answers is not the same as actually finding those genes. In other kinds of medical studies, when relevant genes are in fact isolated and confirmed, they are generally determined to be no more than 10% responsible for the existence of a trait. That's a whole lot less than 77%. "There are no biological determinants for narcissism," says Kagan flatly, "none that are known, at least. I see the literature every day, and no scientific paper worth reading has shown that people who are considered narcissists share some brain profile or set of genes."

That will come as no news to the Freudians. They have long posited two competing theories to explain the narcissistic personality, and neither is genetics-based. The first, put forth by Freud himself, is known as the mask model—the idea that narcissism is just a pose to conceal its opposite: deep feelings of worthlessness and cripplingly low self-esteem, typically the result of poor parenting. "Narcissistic personality disorder is likely to develop if parents are neglectful, devaluing or unempathetic to the child," wrote famed psychologist Heinz Kohut, who was born in Austria in 1913, during the hot-

QUICK, WHAT'S MY NAME? *When Donald Trump puts up a building, he makes sure no one will have any doubt who built it.*

house years of Freudianism. "This individual will be perpetually searching for affirmation of an idealized and grandiose sense of self."

The alternative theory suggests that the adult narcissist is actually paying the price for too much parental praise. In this case, a childhood spent being adored and applauded makes it impossible to adjust to a harsher world in which laurels are harder to come by. The leader of this more direct school of thought was—and in some ways still is—psychologist Theodore Millon, an emeritus professor at Harvard, who maybe did more than anyone else to frame the clinical definitions of personality disorders in the 1970s. "Narcissistic personality disorder comes from unrealistic parental overvaluation," he wrote. "Parents pamper and indulge their youngsters in ways that teach them that their every wish is a command, that they can receive without giving in return, and that

they deserve prominence without even minimal effort."

Then again, it's possible that both the too-much and too-little configurations can cultivate narcissism. In the same way that both a day spent hard at work and a day spent lying around can leave you feeling exhausted, so may these two radically different upbringings lead to the same unhappy place.

Another key driver of narcissism, of course, is culture—and it's been doing its part for centuries. The liberalization of the church in the 16th century, the rise of capitalism, the pursuit of private wealth, and the birth of democracies built not on fealty to a king but on the rights of individuals have all over time elevated the person over the people, the one over the many. Like the dark matter that holds the universe together and the dark energy that steadily pulls it apart, our communal and individual impulses are forever in tension. And as in the

A Modest Advantage

HUMILITY ISN'T A PARTICULARLY HIP virtue these days, but new research shows that a humble friend is a better friend. "Compassion is hard if you don't have humility," says University of Maine psychologist Jordan LaBouff.

Evolution suggests that subsuming one's needs is a trait likely to be preserved only in species for which cooperation is necessary for survival. LaBouff and his team looked for evidence of this humility-helpfulness connection in a study published in *The Journal of Positive Psychology*. "We defined humility as being relatively down to earth and capable of understanding one's own strengths and weaknesses appropriately, not underestimating or overestimating them," says LaBouff. "It's not a meek 'I'm no good,' low-self-esteem feeling." They also framed it as basically the opposite of arrogance or narcissism, using a questionnaire that measures these traits with responses to statements such as "Some people would say I have an overinflated ego" or "I am an ordinary person who is no better than others."

In the first of three experiments, college students were surveyed online. In addition to the researchers' humility test, they also took personality tests and reported on their level of various volunteer and altruistic activities. Of all the traits measured, humility was the most strongly linked with helpfulness.

Problem is, it's hard to rely on self-reporting of helpfulness and humility: arrogant people may try

to appear humble; humble people may downplay their actions; and all of us may take credit for more good works than we actually do. "If I ask people whether they're humble and they say yes, that's tricky to interpret," LaBouff says. To counter this bias, the team measured humility in another group of students with an implicit association test (IAT), which gauges attitudes by examining reaction time in word tasks. For example, if a person is slower to link scientific or mathematical words to women than to men, it may suggest sexist leanings. Similarly, if you consider yourself humble, you're likely to connect more quickly to words associated with humility than someone who defines himself differently.

Researchers found that those who rated high on implicit humility were willing to give 43% more time to help a student in need than those who scored low. Overall, humility explained about 10% of the variability in helpful behavior, LaBouff says, calling that number significant given all the factors that go into deciding to help or not.

In a final experiment, the researchers looked at whether peer pressure affected willingness to help. This time, students listened to a purported campus radio broadcast about a woman whose parents and sister had died and now was struggling to stay in school while supporting her surviving siblings. Each subject was asked to sign up to help the woman on a sheet that had apparently been filled out by seven prior participants. In half the cases, five of the seven participants were said to have volunteered, making it seem as if "everyone else was doing it" and that it would be unacceptable not to. In the other half, only two of the seven students had volunteered.

Seventy-seven percent of those who rated highest on implicit humility volunteered when few others did, compared with 48% of those rated lowest. In the face of universe, which is slowly expanding, the dark energy of our me-first impulses is winning.

"If you take just the last 600 years of human history, we've gone from communal, religious societies to individualistic, secular ones," says Kagan. "First comes Martin Luther, then Adam Smith, then the Declaration of Independence, then diverse societies like in the U.S. and Europe, where nobody can agree on the same set of values." Take that individualistic history and turbocharge it with our modern self-loving, self-promoting, self-esteem-touting value system, and it's no wonder we've gotten plain drunk on ourselves.

Many people do obviously grow beyond their self-absorption, though, even if they have to move well into adulthood before they do. The brain's prefrontal cortex is not even fully wired until we're in our late 20s or early 30s, a time that often brings a new kind of perspective and humility. Life experiences can be sobering too. A few setbacks or disappointments—to say nothing of the social opprobrium that comes from caring far too much about yourself and far too little about everyone else—can temper even the most extreme behavior. Inevitably, there are those who won't ever take the hint. For them the road to growth has to run through a psychologist's office, where they can learn new ways to see the world and their place in it. That process takes work and patience, not to mention a willingness to defer to a wiser mind, neither of which narcissists do easily.

The alternative, however, is far less pleasant. A life devoted to seeking out an admiring audience—particularly from a pool that finds you less and less appealing—rarely ends well. There's a lot to be said for hearing the applause. But there's just as much to like about offering it up to others.

DOWN TO EARTH *The link between humility (such as that of the Sisters of Mother Teresa) and helpfulness toward others is strong.*

greater perceived social pressure, though, the humble and less humble tended to have the same urge to help. "Americans overwhelmingly say that they value humility," LaBouff says. "They want their friends to be humble and they say they want to be humble, but expressions of it tend to be rare."

Indeed, in a culture that increasingly rewards self-promotion and essentially punishes those who won't do it, humility seems endangered. LaBouff does cite earlier research suggesting that humble people actually tend to outperform egoists in organizational hierarchies. One survey of CEOs and middle managers found that humility was an important component of effective leadership.

However, displaying humility has proved a double-edged sword for anyone besides older white men. Studies have shown that when youth, women, or minorities admitted errors or doled out credit, their competence was questioned. White men, in contrast, were rewarded for those behaviors. A larger follow-up study revealed that humble leaders produce more engaged employees and less voluntary turnover. Ironically, though, LaBouff notes, humble employees had a harder time advancing. —*By Maia Szalavitz*

Hijacked by Worry

Few things imprison a mind quite like obsessive-compulsive disorder, but better treatments are breaking its hold.

BY JEFFREY KLUGER

SAY YOU LEAVE WORK AT 6 p.m. for your 12-minute drive home. Say just as you pull onto the street, a child on a bicycle crosses in front of you. A few feet later, you feel the thump of a pothole. But was it a pothole? Did you hit the child? You check the rearview mirror. All clear. Are you sure? You circle back around. Still clear—except for a bag of leaves on the curb. Wait—is it a bag or a child? You circle again. Four hours later you arrive home, mutter to your spouse about a late meeting, and go to bed spent and ashamed. Tomorrow you'll do it all over again.

That is the serious mental glitch known as obsessive-compulsive order (OCD), and devoting an entire evening to a 12-minute drive is not the only way to know you've got it. You know it when you shrink from the sight of a kitchen knife, worried that you'll inexplicably grab it and hurt yourself or a family member. You know it when leaving the house consumes hours of your day because the pillows on your bed must be placed just right. You know it when you can't leave the house at all for fear of a vast and vague contamination you can't even name.

We think we can recognize OCD. It's the guy from *Monk*; it's Jack Nicholson in *As Good as It Gets*. In the end, though, things work out for them. They even get the girls, who see them as adorable emotional fixer-uppers.

But OCD isn't adorable. About 7 million Americans are thought to have one form or another. Worse, a single case can destabilize a household, claiming several collateral victims. Confusing the issue, OCD often masquerades as depression, bipolar disorder, attention-deficit/hyperactivity disorder (ADHD), autism, even schizophrenia. And sufferers often conceal their problem, ensuring that no diagnosis—right or wrong—can be made.

In fact, the average lag between onset and diagnosis is a shocking nine years, according to surveys of doctors conducted by the International OCD Foundation. It takes an average of eight more years before effective treatment is prescribed. If the disorder strikes early, as it often does, an entire childhood can be lost. "OCD had a slow research start," says Dr. Gerald Nestadt, director of the OCD clinic at Johns Hopkins University. "It's behind schizophrenia, bipolar disorder, autism, and ADHD."

But now a burst of genetics studies is turning up clues to the causes of OCD. Scanning technologies are pinpointing the parts of the brain that trigger symptoms. New treatments are being developed, and refinements of older ones, like talk and behavioral therapy, are proving effective. "We all have intrusive thoughts, but most of us consider them meaningless and move on," says psychologist Sabine Wilhelm, associate professor at Harvard

Medical School and director of the OCD program at Massachusetts General Hospital. "In people with OCD, the thoughts become their lives. We can give their lives back."

A little anxiety is a good thing. It was not enough for early humans to know there was no lion near the family cave; they had to be urged to imagine all the other places a lion could lurk. Anxiety about the ways harm may befall someone else keeps us mindful of the safety of family and community. "There's a creative, what-if quality to this thinking," says clinical psychologist Jonathan Grayson of the Anxiety & OCD Treatment Center in Philadelphia. "It's evolutionarily valuable."

Something woven so tightly into the genome isn't about to be shaken loose by a few thousand years of modern living. But it doesn't mean everyone with eccentric traits—the woman in the next office who gets edgy when something is moved on her impeccably neat desk, for example—has OCD. "Having OCD-like traits is a universal experience," says Judith Rapoport, author of the landmark book *The Boy Who Couldn't Stop Washing* and chief of child psychiatry at the National Institute of Mental Health. "I sometimes count on my fingers when I have nothing to count." The key to diagnosing whether such behavior is OCD is how great an impact it has on your life. "You have to show longstanding interference with function, and that eliminates most people," Rapoport says.

HOWIE MANDEL
The comedian and TV host kept his condition secret for many years. He bumps fists to avoid shaking hands, never touches a handrail, and wears a surgical mask when his kids are sick.

Most Common Expressions of OCD

RELATIONSHIP SUBSTANTIATION
A compulsive search for tiny but disqualifying flaws in a partner or spouse. Romances and marriages often do not survive the scrutiny.

FEAR OF INJURING OTHER PEOPLE
A preoccupation with the idea of losing control and injuring or even killing others, it often results in a terrified avoidance of knives, scissors, or other sharp objects.

RESPONSIBILITY ANXIETY
A broader fear of negligently hurting others. Sufferers will smooth out throw rugs or pick up trash from sidewalks so strangers won't trip.

SCRUPULOSITY
An intolerance of disorder or asymmetry, this is a fastidiousness that goes way beyond tidiness. One of the better-known forms.

CONTAMINATION ANXIETY
The familiar hand-washing compulsion—but that's only the tip of it. Feared contamination can spread from hands to other objects; that can lead to clothes, belongings, and even walls being washed.

SEXUAL-ORIENTATION FEARS
People who may have no moral or social objections to homosexuality become consumed with a fear of discovering stirrings of it in themselves. It's perfect OCD fuel, as it's impossible to prove the negative.

HYPOCHONDRIA
This was previously seen as a freestanding condition, but is now widely thought to be a form of OCD. Sufferers may fixate on the fear that they have a particular disease or a changing series of them.

What causes only some to suffer that interference? Why does their internal alarm keep shouting "lion!" long after they've checked everywhere that lion could be? The answer always figured to lie principally in a small, almond-shaped structure in the brain called the amygdala, where danger is processed and evaluated. If this risk center is overactive, the thinking went, maybe it would keep alerting you to peril even after you've attended to the problem.

As it turns out, though the amygdala is a big player in the process, it's only one of several. Functional magnetic resonance imaging (fMRI) and other scanning technologies have allowed researchers to peer deeper into the OCD-tossed brain. In addition to the amygdala, three other anatomical hotspots are involved: the orbital frontal cortex, the caudate nucleus, and the thalamus—the first two seated high in the brain, the third lying deeper within. "Those areas are linked along a circuit," says Dr. Sanjaya Saxena, director of the OCD research program at the University of California, San Diego. "And the circuit is abnormally active in people with OCD."

But while scans can describe the landscape of the obsessive-compulsive brain, they can't tell you how it got that way. As with many psychological disorders, OCD has a powerful genetic component. A blood relative with the condition puts your risk at 12%; if that seems low, it's more than four times higher than the U.S. population as a whole. Drilling down, researchers at Johns Hopkins found six gene markers that appear more frequently in kids with the disorder than in unaffected people. Another group identified a seventh gene that appears to regulate a brain chemical known as glutamate, which stimulates signaling among neurons. Too much glutamate, and the signals keep coming. In the brain's alarm centers, that means the warning bell rings and rings.

What makes the glutamate-related gene especially suspect is the people it affects most. OCD strikes males and females about evenly, but early-onset forms target boys more than girls, especially boys who also exhibit the involuntary tics or vocalizations of Tourette syndrome. Interacting with the glutamate gene are three others related to androgens, or masculinizing hormones. Interacting with those is another that has been implicated in Tourette. Gathered in the same chromosomal neighborhood, they can make real trouble. "Kids who start early tend to be boys, tend to have tic disorders, and tend to have parents with tic disorders too," says Dr. John March, retired chief of child psychiatry at Duke University.

Other compelling, if controversial, research has pursued an entirely different cause of OCD: streptococcal infection. As long ago as the 17th century, British physician Thomas Sydenham noticed a link between childhood strep and the later onset of a tic condition that became known as Sydenham's chorea. Modern researchers who see a link between tics and OCD also wonder if, in some cases, strep might be involved with both.

In 2007, investigators from the University of Chicago and the University of Washington studied a group of 144 children—71% of them boys—who had tics or OCD. All the kids, it turned out, were more than twice as likely as others to have had a strep infection in the previous three months. For those with Tourette symptoms, the strep incidence was a whopping 13 times as great.

The tics and OCD are likely the result of an autoimmune response, in which the body attacks its own healthy tissue. Blood tests of kids with strep-related tics and OCD have turned up antibodies hostile to neural tissue, particularly in the brain's caudate nucleus and putamen, regions associated with reinforcement learning. "There certainly seems to be an epidemiological relationship," says Dr. Cathy Budman, director of the movement-disorders center at North Shore University Hospital in Manhasset, N.Y., "but what it means needs further investigation."

No matter how or when the disorder hits, a common first step in striking back is short-term behavioral therapy, specifically a technique known as exposure and response prevention (ERP), in which sufferers actually seek out the cause of their anxiety. Eventually, emotional nerve endings grow desensitized to the stimulus. The goal is to tough it out until then.

At a recent Obsessive Compulsive Foundation convention, Grayson, the Philadelphia-based psychologist, gave those with OCD a quick taste of ERP. He invited audience members with dirt and germ anxieties to sit beside him on the ballroom carpet, then told them to touch the carpet and bring their fingers to their lips. Left to themselves, most would have refused or, if they went along, would have then found the nearest bathroom and spent long minutes—perhaps hours—scrubbing. Instead, they sat with their anxiety, learning a first lesson about how pain subsides. Extended ERP treatment involves a gradually increasing series of exposures.

Such tactical jujitsu works for all manner of OCD, though it's not always easy to find a doctor skilled at administering it. Patients obsessed about their sexual orientation, who become intolerably anxious if they so much as notice an attractive member of the same sex, are assigned to flip through magazines for scantily clad same-sex models. People plagued by what's known as relationship substantiation, who are consumed by inconsequential defects in a partner, are encouraged to seek out those flaws and even exaggerate them in their mind.

Medication helps too. Antidepressants such as Prozac and other selective serotonin reuptake inhibitors can dial down the anxiety enough that patients can get started with ERP and, significantly, stay with it. When patients are children, doctors are reluctant to prescribe medication, but they are also careful not to stay too long with ERP alone if it's not producing results. "The longer a child struggles with the illness, the more impact it's going to have," says Dr. John Piacentini, director of UCLA's child-OCD clinic. Some acute cases of OCD require more extreme methods, such as hospitalization, intensive exposure therapy, and stronger medications.

Dr. Vladimir Coric, past director of OCD research at Yale University, is among a growing group testing drugs that target the glutamate problem. The best option so far, riluzole, was first developed for Lou Gehrig's disease and works by turning down the glutamate spigot. In Coric's admittedly small studies and clinical observations, half of about 50 subjects experienced at least a 35% remission, and almost all the rest improved at least a little.

Much more invasively, some clinicians are turning to deep-brain stimulation (DBS), in which electrodes are implanted in the brain and connected by wires embedded in the skin to a pacemaker-like device in the chest. Low doses of current are applied as needed to calm the turmoil in the regions that cause OCD. The procedure sounds extreme—and it is—but it's already been used in over 40,000 people worldwide to treat Parkinson's disease. "Many of our OCD patients are able to reengage in life rather than being stuck at home," says Dr. Ali Rezai, director of the neuroscience program at Ohio State University Wexner Medical Center, who performs DBS procedures.

For the vast majority of victims, treatment need never go so far. OCD, for all the suffering it inflicts, is nothing more than the brain doing something it's supposed to do—warn you of danger—very badly. Living in the world means living with risks: real, imagined, or exaggerated. That's not an easy lesson, but it's a powerful one. And once learned, it can offer a paradoxical state of peace.
—*With reporting by Dan Cray/Los Angeles and Rachel Pomerance/Atlanta*

The Case for Labeling

Many people worry that a psychological diagnosis marks you for life.
This writer wishes she'd been so lucky.

BY MAIA SZALAVITZ

When I was a toddler, I hated to be held.

I couldn't tolerate itchy fabrics or new foods. I had intense intellectual obsessions, which I went on about ceaselessly, and I took teasing literally, which, of course, made me a magnet for bullies. Although no one taught me, I could read by the time I was 3.

I almost certainly met the criteria for the now much-debated Asperger's syndrome—a diagnosis that has gone from being vanishingly rare to being part of an autism spectrum that is understood to affect about 1% of children. Asperger's is far from the only developmental disorder or psychiatric condition with an incidence that has increased dramatically in children in recent years. Attention-deficit/hyperactivity disorder, depression, bipolar disorder … you name it, it's spreading.

Some argue that this speaks to a society-wide rage for labeling, that we can't wait to pathologize what is really just normal childhood awkwardness, sadness, or rambunctiousness. Others say science has gotten better at detecting very real problems. My experience suggests that this debate is missing a key perspective. The most important issue is not how a child ends up with a label, but how it affects her if she does.

Because I was born in the 1960s, I was never diagnosed. The anti-label caucus would say I was lucky, that I was saved from a stigmatizing burden and my parents from seeing my prospects as limited. But I found not being diagnosed to be a very mixed blessing. I still had to live with nearly all the behaviors of Asperger's, but I had no help in understanding them. Instead of being psychiatrically labeled, I was morally tagged. Relatives and teachers called me "bossy" for my obsessiveness and the rigidity with which I tried to manage my sensory overreactions. I was "selfish" for not listening to others. I was "not a people person" because of my obsession with ideas and difficulty making friends.

As a result, the characteristics now associated with the autism spectrum felt like character defects. If I was intellectually "gifted," I also felt cursed because so many of my other traits colored me as a bad person. Not knowing that there was a relationship between my talents and my strange compulsions, I just saw my inability to connect as hopeless.

Indeed, I came to believe that the only way I'd ever relieve my loneliness was to become incredibly successful; people would like me if there was something they could get from me. Problem was, my drive and the overwhelming self-hatred behind it—exacerbated by relentless middle-school bullying—coalesced into depression and eventually addiction once I sought support in the drug culture in high school. In college, the social dislocation destabilized me further. Continuing to seek refuge in drugs, I became deeply involved with cocaine. I'll skip the disaster that came next: freebasing, shooting heroin, rehab, and finally recovery. But I will say that had I been diagnosed, it might have given my life a very different outline.

If I had known there were others like me, that I wasn't "bad" because I was oversensitive or needed things to be a certain way, and that I could learn to socialize better, I might not have hated myself so much. Freed from self-loathing, I might not have sought relief in drugs.

Maia Szalavitz is the author of Help at Any Cost: How the Troubled-Teen Industry Cons Parents and Hurts Kids.

Is it possible that I would have seen an Asperger's label as limiting, or its own cause for despair? Sure. I could have used it as an excuse—to not try socially, to withdraw. Like some other children with Asperger's, I might have found those oft-prescribed social-skills classes demeaning and useless.

But I don't think so. It's certainly not how I responded when, at 23, I accepted another label. When I finally admitted I was a drug addict, I was actually comforted to realize I was suffering from a medical disorder that entire communities were devoted to healing. Embracing the addict identity was liberating because it let me know where I stood and, just as important, let me see that others like me had gotten better. Yes, there was a stigma attached—but it was something to fight, not something to avoid.

Of course, labeling children does present some very real issues. For starters, developmental trajectories are difficult to predict. No one knows which teen drug user, for example, is embarking on a life of addiction and which teen will easily refrain when family or work responsibilities kick in. Telling young marijuana smokers in rehab that they have a "chronic, lifelong disease with a 90% chance of relapse" carries the risk of becoming a self-fulfilling prophecy. Similarly, forcing teenagers to come to grips with any label—particularly a pathological one—can do emotional harm at a time when their identity is particularly vulnerable. And sometimes it seems as if we have just moved from one extreme (childhood misbehavior is "bad") to another (childhood misbehavior is "mad") without finding that middle ground where reality lies.

But if we want to give children the best chance to develop their best selves, we have to help them navigate some pretty tough territory, balancing the need to take responsibility for their actions with an understanding of what drives their behavior and how their unique needs can be met. And it is labels that will guide them—to explain, not excuse; to influence, not predestine.

When I tell people I would be diagnosed with Asperger's if I were a child today, they typically disbelieve me. That's because these days I'm perceived as warm and supportive, two traits that the uninformed still think are incompatible with my condition. The stigma is still very real.

I make more sense to some people if they think I've somehow "outgrown" Asperger's, or that I had only a mild case to begin with. All I can say to them is that this "mild" case continues to affect my behavior many times every day and has influenced virtually every major decision I've ever made. I might also inform them that research suggests that girls with Asperger's are at greatest risk not of overdiagnosis but of being overlooked. Because social pressure on females is even more intense than it is for males, it is far easier for us to slip under the radar, as our obsessions tend to be different from those seen in boys. Some now theorize that anorexia—with its rigidity, compulsiveness, and particular obsessive focus—may in some cases be a predominantly female form of autism.

You can call what I have whatever you want, as long as you call it something. Because not having a name for it profoundly affected my life. Diagnostic labeling that's incorrect or stigmatizing can do harm, but the labels that undiagnosed children give themselves (or are given by others) are far more dangerous.

Dynamics

Why eligible singles shouldn't take those online dating questionnaires quite so seriously; why we should all get a good night's sleep; and why our children should put down their screens—at least sometimes.

What Makes Us Get Along?

Dating sites reduce it to an algorithm, but science is still probing nature's rules of attraction.

BY BELINDA LUSCOMBE

TIM BERNERS-LEE, THE PERSON MOST RESPONSIBLE FOR THE CRE-ation of the World Wide Web—and therefore a guy who knows a thing or two about making connections—likes to talk about the Stretch Friend. "A Stretch Friend is somebody who is like a university that might be a bit hard for you, but if you can get in, it would be great," he says. One of the strengths of the Web, Berners-Lee believes, is that it fosters Stretch Friends—very different people who normally would not be within each other's reach.

His scenario is plausible enough. The Internet does allow users to meet on neutral ground where they can take the time to discover what they have in common. Opposites would inevitably attract. Everyone would find the other who completes them.

But human nature laughs at logic. The short history of digital communication proves pretty conclusively that given the choice to hang with anyone in the world, we tend to cluster in like-minded groups, connecting most energetically with those whose hobbies, politics, and appetite for wittily captioned photos of cats match our own. And if that reality does seem a bit limiting—are our best friends destined to be those who are most like us?—it, too, makes good sense.

Why we like who we like seems both impossibly random and self-evidently simple, which is just another way of acknowledging that nobody knows for sure what draws us to this person but not that one. How big a role does personality—ours and theirs—play in

Brothers From Another Mother
MICK JAGGER AND KEITH RICHARDS
For 50 years they've caroused, bickered, and written more than their share of rock's iconic songs.

Powder Kegs
ELIZABETH TAYLOR AND RICHARD BURTON
Despite operatic fights, drinking binges, and two divorces, they are still considered one of showbiz's greatest love stories.

choosing friends and lovers? Is there a particular type of partner for each of us? How do people who seem very different manage to get along anyway? And do we even have a choice in this most fundamental matter? These questions occupy the minds of managers, sociologists, and, of course, anyone who has ever created an online dating questionnaire. As well they should: we all have a vested interest in figuring out how to make families, workplaces, and communities more functional.

At the heart of how we relate to one another is the thorny issue of whether we even want what we think we want. Take the small but telling study of a campus mixer conducted at Columbia University in 2007. The point of a mixer, is to, well, mix, to meet new people, make connections, cross-pollinate ideas. Toward that end, all the attendees stated beforehand that they wanted to extend their network and open themselves to different experiences. But their special tracking name tags told the opposite story. Mostly, partygoers talked to people who were pretty much just like them.

A wider study, conducted at the University of Kansas in 2011, confirmed that result—with an interesting bonus. Researchers explored the friendship patterns at two types of schools: the 25,000-student university campus in Lawrence, and four significantly smaller colleges (average enrollment: about 1,000) scattered throughout the state. Reason suggests that students at the larger—and almost by definition more diverse—school would gather a more diverse group of friends. In fact, though, the kids in Lawrence tended to befriend those who mirrored their own beliefs, values, attitudes, and personalities more closely than did those on the more self-contained campuses. Meanwhile, students at the small colleges seemed to have closer relationships, even if a pair of friends were less similar. Only when opposites are given few other choices, the study seems to conclude, do they attract—but when they do, that attraction seems to be strong.

The desire to seek out people like us extends to the workplace. Ask managers if they'd rather have a team with similar or disparate skills and personalities, and they'll almost always argue for the latter. But a 2012 study of hiring techniques at some of the most demanding workplaces—banks, law firms, management consultancies—found that, actually, people were hired as much for their likability as their skill sets. And what made them likable to interviewers? Common interests. As the researcher Lauren Rivera put it, "Employers sought candidates who were not only competent but also culturally similar to themselves."

If like really does appeal to like, what does that mean for that most important life decision? Do we end up settling down with mirror images? If only it were so easy.

One of the unexpected byproducts of the massive online dating industry is that scientists have been blessed with tons of data regarding what people look for in a mate. And as they begin to rummage through it all, they

Masters of the Universe
BARACK AND MICHELLE OBAMA
The most influential couple in the world makes time for PDAs—and dinner at home with their daughters.

have already hit on some pretty interesting conclusions. Biological anthropologist Helen Fisher, a professor at Rutgers University who authored the book *Why Him? Why Her?*, has found evidence to suggest that compatibility is associated with temperament. "There are two parts of personality," she explains. "Character, which is everything you grew up to believe and do and think, and temperament, which is your inherited traits. What I'm trying to do is add the role of biology, of temperament, to our understanding of love." Working backward from what she found in the academic literature and a study of more than 40,000 people on a dating website (she's a paid consultant for Chemistry.com), Fisher isolated four temperament types. Each correlated more or less with one of the body's neurochemicals.

Explorers, the first group, express dopamine and tend to be risk-taking, curious, creative, impulsive, optimistic, and energetic. Builders, associated with the serotonin system, are cautious but not fearful, calm, traditional, community-oriented, persistent, and loyal. Directors have traits associated with testosterone: they tend to be analytical, decisive, and tough-minded. The final group, Negotiators, manifesting traits associated with the es-

trogen system, are broad-minded, imaginative, compassionate, intuitive, nurturing, and idealistic. According to Fisher's calculations, Explorers like Explorers (think Brad Pitt and Angelina Jolie) and Builders like Builders (George and Barbara Bush), while Negotiators and Directors are drawn to each other (Denis and Margaret Thatcher—or Phil and Claire from *Modern Family*).

Fisher has also noticed in her three-year analysis of the data that personality has supplanted more traditional factors—religion, politics, ethnicity—as the one most important to daters. This is not an industry-wide finding, however. Not surprisingly, recent data from Spark Networks, which runs the religion-oriented websites JDate and ChristianMingle, indicate that almost 75% of those surveyed are in the market for a same-faith match, though a different religion was an absolute deal breaker

Besties
AMY POEHLER AND TINA FEY
BFFs since 1993, they've conquered the comedic universe with their sitcoms, movies, and parodies of Sarah and Hillary.

for only 16%. (Bonus tidbit: Jews favored intelligence in a potential mate; Christians preferred a sense of humor.)

That Fisher and Spark Networks have come to different conclusions wouldn't at all shock the many researchers who are skeptical about whether online dating data can even be trusted. At root, their qualms have to do with how important a role personality plays in attraction. "There's surprisingly little work showing that similarity or dissimilarity in personality is important in predicting relationship outcomes," says Eli Finkel, professor of psychology at Northwestern University, whose 2012 study threw doubt on the efficacy of online dating sites. "At the end of the day, there's very little evidence that opposites, or similar types, attract."

Some psychologists think that evolutionary urges are the primary determinant of successful hookups. Men desire women who look good to them because attractiveness implies a healthy fertility that assures the men that their DNA will survive another generation. Women prowl for providers for the same reason, a guarantee that their children will be taken care of. Nothing trumps these primal needs, according to the father of human-mate-choice studies, David M. Buss, a professor of psychology at the University of Texas.

In addition, this faction of researchers is loath to put too much stock in dating questionnaires because why we love—or even like—is not always a clear-eyed proposition. After all, people have been known to fall for someone who upends their ideal. It's not so much that daters lie about what they want as that they are blind to how much romantic choice is affected by circumstance, not to mention Cupid's arrow—more technically known as the blend of neurochemicals that spurs infatuation.

"What people are looking for, you just can't ask them," says Peter Todd, a professor of cognitive science, informatics, and psychology at Indiana University. Todd conducted a nifty study in which he asked people to list what they were looking for in a mate, then watched them participate in a speed-dating event in Germany to see how closely their preferences were reflected in the partners they were drawn to. Although most of the speed daters mentioned attributes that reflected their own, they actually selected partners more in line with what evolution might suggest, had it been sponsoring the event. "Men chose women based on physical attractiveness, and the women, who were generally more discriminating, chose men whose overall desirability matched the women's self-perceived physical attractiveness," says Todd.

In the end, some experts say, we may all just be looking for the wrong thing. "It turns out there's a lot more research supporting a different thinking about what makes people get along well," says researcher and couples therapist Melissa Schneider. "It's called relationship aptitude. Aptitude is made up of individual qualities, which includes personality but also attachment style, mental-health profile, and experiences like parental abuse, child

For the Love of the Art
FRED ASTAIRE AND GINGER ROGERS
She said their working relationship was "cordial, if distant," but on screen they were the epitome of elegance and glamour.

abuse—things that shaped you." Put another way, some personalities are more suitable matches not to any other particular personality but to other humans in general.

In the search for a soulmate, these psychologists say, the word to hunt by is OCEAN, the acronym for the Big Five personality traits: openness, conscientiousness, extroversion, agreeableness, and neuroticism. Marriages—or any partnership, really—in which both parties have a healthy amount of the first four traits and hardly any of the fifth tend to fare better. No matter how alike or complementary two humans are, if either struggles with emotional stability, a successful connection is a long shot. "The trait most associated with relationship well-being is neuroticism—a tendency to experience negative and volatile emotions and not manage environmental stress," says Finkel. "It's a very strong predictor of problems in marriage."

Agreeability, too, has an outsize effect. "Those who are agreeable tend to report happier relationships—and their partners agree," says Jeff Simpson, professor of psychology at the University of Minnesota and vice president of the International Association for Relationship Research. But, Simpson notes, spouse seekers who are highly emotionally charged or disagreeable need not face a life of loneliness. "If you find a partner who is good at regulating emotions and at making you feel you can trust them," he says, "they can help suppress the expression of negativity."

Then again, as in attraction itself, the study of it may be colored by the eye of the beholder. John Hibbing is a professor of political science at the University of Nebraska. After looking at data culled from thousands of marriages in the United States, he didn't find much in common between partners in terms of physical attributes and personality traits. But guess where he did find a correlation? Right. "Political attitudes display interspousal correlations that are among the strongest of all social and biometric traits," he wrote in a 2011 paper. Apparently, couples who caucus together stay together—and not only after decades of hashing out differences at the dinner table. "It appears the political similarity of spouses derives in part from initial mate choice rather than persuasion and accommodation over the life of the relationship," says Hibbing. He doesn't suggest we interrogate each other about our voting history over that first get-to-know-you cocktail; it's not actually party affiliation that's behind the more successful bondings. "Political views are really based very deeply in our psychology, tastes, preferences, and biology," Hibbing explains. "It kind of oozes out from us—people have a general sense of our values from what we do. Politics is kind of a general values structure that we look for in a mate."

Which kind of sounds as if Hibbing thinks personality does matter after all, doesn't it? Clearly, the science of attraction is its own kind of coy flirt, as every advance is countered by one more obstacle. Here's another: why do identical twins, who often like the same clothes and décor and even have many of the same interests and values, go for such different mate choices? Studies show that spouses of identical twins have no more chance of being similar to each other than spouses of fraternal twins do, and only slightly more chance than spouses of randomly selected adults.

With so much conflicting—or at least not exactly confirming—information swirling around them, what are today's searching singles to do? Just what they've always done, in all the ways they've always done it. In the end, almost all researchers agree, what's most important is not who you marry, but how you are married. Although there is nothing more romantic than the notion of a soulmate, the one perfect physical and psychological match, it can be destructive too. People who think their work is over once they've located Mr. or Ms. Right are invariably in for a disappointment. And they are more likely to walk away when problems arrive, rationalizing the breakup by assuming that they chose wrong. "On the other hand," says Finkel, "people with strong beliefs in romantic growth—that happy relationships emerge from overcoming challenges—are especially likely to persist and succeed when confronted with problems." Getting along, then, has a lot less to do with that one big decision—who you choose—and much more to do with all the little choices that come after.

Hot and Cool
JOHNNY AND JUNE CARTER CASH
He was a notorious hellraiser, she a fundamentalist Christian, but the country-music legends stayed married for 35 years.

Superstars
VICTORIA AND DAVID BECKHAM
The Spice Girl and soccer icon, married 13 years, bask in the limelight and often coordinate their fashion-forward ensembles.

Shhh!
Genius at Work

*New research explains why your brain may be
at its most creative when you're asleep.*

BY JEFFREY KLUGER

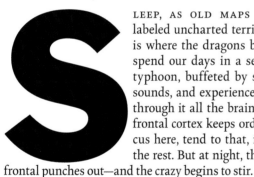

SLEEP, AS OLD MAPS ONCE labeled uncharted territories, is where the dragons be. We spend our days in a sensory typhoon, buffeted by sights, sounds, and experiences, and through it all the brain's prefrontal cortex keeps order: focus here, tend to that, ignore the rest. But at night, the prefrontal punches out—and the crazy begins to stir.

To sleep is to enter a world entirely like our own, and entirely unlike it too. You can board a plane that's really a car that flies to Russia, except it's the moon and your mom is there, until she's your dad. Dreams can be prosaic or repetitive—how many times can you show up at the same party in your underwear before you remember to put something on?—but whatever they are, they remain mysterious. As a sleeping brain runs its absurdist-movie loop all night long, it has always taken care to conceal the mechanisms behind it.

But now neuroscientists are getting a peek behind the screen. A growing arsenal of tools—fMRIs, PET scans, high-density EEGs—allow them to see the brain at work throughout the sleep cycle. And that is helping to explain one of the mind's most ineffable qualities: creativity.

We've all slept on a problem and had it sort itself out by morning. But that's just one thing the brain on nighttime autopilot can do. Paul McCartney came up with the melody for "Yesterday" in a dream; Elias Howe, the inventor of the sewing machine, is said to have solved his needle problem when he dreamed of an attack by warriors carrying spears with holes in the tips. "Dreams are just thinking in a different biochemical state," says Harvard University psychologist Deirdre Barrett, author of *The Committee of Sleep.* "In the sleep state, the brain thinks much more visually and intuitively."

The hunt for the source of human creativity has gone on as long as people have been creating. We want to know how celebrated inventors came up with the next big thing and how we'll find our own brainstorm when we need it. It's no secret that sleep can be a well of good ideas. We're finally learning how to dip into it.

The act of sleeping, as researchers have long known, entails a lot more than simply conking out for the night. There are two main sleep cycles—rapid eye movement (REM) and nonrapid eye movement (NREM)—which alternate through the night. But eye movement is a fraction of what defines the phases. NREM sleep starts as a doze, no greater than snorkeling depth, that steadily deepens to levels in which muscles relax, heart rate and respiration slow, and body temperature drops.

REM sleep, which begins about 90 minutes after the first NREM cycle, is the true blue ocean of sleep. Heart

MARLON BRANDO

RICHARD NIXON

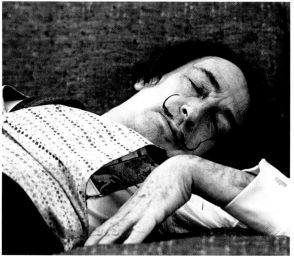

SALVADOR DALI

THOMAS EDISON

rate and respiration accelerate; brain activity, as measured by electroencephalograms (EEGs), increases too. All are functions of dreaming. Muscles become paralyzed, lest you act out the scenes unspooling in your head. Those dreams in which you're trying to run away but can't seem to move your legs are no illusion.

Most REM sleep comes in the last four hours of the night, says cognitive neuroscientist Jessica Payne of the University of Notre Dame. "Dreams in the early, NREM phase can be kind of literal. It's in the REM phase that you get all these crazy binding errors—you dream that you're walking in Peoria and suddenly wind up in Paris."

"Binding error" is the rare scientific term that means pretty much what it sounds like. Waking brains are orderly, but sleeping brains are fragmented, and those bits can be reassembled the wrong way. In this case, though, "wrong way" means untried avenues worth exploring.

In a frequently cited 2009 study, investigators at UCLA and the University of California, San Diego, asked a group of volunteers to solve a word puzzle known as a

remote-association test. In a typical question, subjects are given three words and asked to determine a fourth that links them all. The answer for "broken, clear, and eye," for example, would be "glass." Volunteers took the test twice; between the two sessions they were told to take a nap. Some just rested, others dozed, and some tumbled into REM sleep. In round two, the scores of those participants who got a slug of REM improved 40%, while the rest of the cohort saw their scores drop. Sleep, it appears, had sharpened the brain's ability to accomplish this particular kind of reasoning.

A study performed at the University of Lübeck in Germany came at the same idea in a more revealing way. Subjects completed a series of tedious math problems that relied on cumbersome algorithms, but hidden within the formulas was an elegant arithmetical shortcut. About 25% of the subjects discovered it on their own. That figure jumped to 59% once they were allowed eight hours of sleep before coming back for more. Subjects who remained awake showed no improvement.

"If you have an idea about a simpler solution and it's been working itself out in your head, you still tend to use the familiar one," says cognitive neuroscientist Howard Nusbaum of the University of Chicago. When you sleep, the better answer gets the chance to emerge.

What gives the brain the ability to utilize its nighttime downtime is something it shares with a computer: the capacity to run multiple programs at once. You can wrestle with a problem even when your conscious attention is elsewhere. The "aha!" when you suddenly remember something three hours later is no accident. "Conscious awareness can focus on one thing at a time," says Barrett. "But problems go on getting processed under the radar."

Sleeping doubles down on this phenomenon. The traffic-cop role that the prefrontal cortex fills when we're awake does more than keep the brain focused on a particular conscious task. It also screens thoughts that it decides shouldn't be thought at all—not just socially inappropriate ones, but rationally inappropriate ones too. In sleep, the brake on imagination is released. That explains the German math study.

While the prefrontal censor dials down, the brain's visual centers, in the occipital lobe at the back of the head, dial up. The hallucinogenic quality of dreams is a result of those centers' mixing images at will. Usually the result is chaff, but not always. One night in 1816, Mary Shelley dreamed of a man assembled from bits beyond the grave. Then she wrote *Frankenstein*.

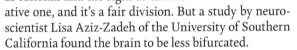

SLEEP SPINOFF
One study has found that 8% of all dreams play a significant role in waking-life creativity.

RAT REHEARSAL
Brain probes show that rats that run mazes during the day practice them mentally when they're asleep.

As important as which regions are on duty is how they communicate. We think of the left hemisphere as rational and the right as the creative one, and it's a fair division. But a study by neuroscientist Lisa Aziz-Zadeh of the University of Southern California found the brain to be less bifurcated.

When architecture students undergoing functional magnetic resonance imaging (fMRI) brain scans were asked to perform a visual-spatial task—arranging shapes in their heads to see if they could assemble them into a square or a triangle—the right, artistic hemisphere carried the load. When the students were given a slightly more creative task—arranging a circle, a C, and an 8 into a face—the right hemisphere tapped the left for assistance. "The regions that are active during the creative process largely depend on the kind of task," says Aziz-Zadeh.

A 2008 study at the University of Rome found a similar process during sleep. With the help of EEGs, investigators tracked communication between hemispheres when subjects were awake, in NREM sleep, and in REM. In waking and NREM states, information traveled mainly from left to right, which is consistent with the idea that the left brain controls the right. During REM sleep, though, there was no preferred direction. The creative right can thus break out of the literal left's shadow.

Synapses—the cell-to-cell links that are the bits of the brain's operating system—play an important role too. Each brain cell can link to more than one other, so you'd think that the more connections, the better, as it makes for a richer system. And it's true to a point. But too many hookups can lead to chaotic free association rather than organized thought. Periodically, the brain needs to clear the synaptic underbrush. It's like "running a repair-and-cleaning program on a computer to defrag the hard drive," says psychologist William Killgore of Harvard Medical School. That cleanup happens during sleep.

The hormone cortisol rises during REM and helps form new and imaginative ideas from the data that survives the defrag. Cortisol, commonly described as a stress hormone, tends to fracture memory. It has the same effect during sleep, and Payne of Notre Dame believes this encourages the unbinding and rebinding of images that define dreams. "The brain dislikes fragmentation, so it weaves narratives," she says. "And that gives rise to novel thinking." Dopamine is another ingredient in the brain's creativity recipe; levels rise in pleasure centers both when we're dreaming and when we're creating, a kind of reward and reinforcement that keeps the dreams—and ideas—flowing.

Can it be, then, that we are all equally imaginative in our sleep, or do people who are more creative in their waking hours retain an edge at night? As nice as it would be to think that sleep is a great democratizer, creative types likely have a round-the-clock advantage. Notre Dame psychologist David Watson found that subjects who scored high on creativity scales when they were awake tended to remember their dreams more. "One reason is that they simply have more vivid and interesting dreams," he says. "That's linked to having an active fantasy life; daytime behavior shades over into night. The rich get richer."

That is not to say that the creative middle class can't aspire to join the metaphorical 1%. The best tactic for remembering dreams is keeping a journal by your bed, says Watson. Also, avoid alcohol and caffeine; both scramble the NREM and REM cycles. Barrett's studies suggest that engaging in some pre-bedtime priming—contemplating a problem you'd like to solve—increases the likelihood that sleep will bring answers. Up to a third of her subjects reported that priming helped lead to a solution that had eluded them during the day.

While those strategies won't guarantee that a good night's sleep will give you all your answers, they won't reduce your favorable odds, either. You have problems every day, after all, and you go to bed every night. Even if you don't think of yourself as creative, your sleeping brain has other ideas.

Amending Your Constitution

We're suckers for stories of personal reinvention, but can people really change the essence of who they are? The answer is transforming.

By David Bjerklie

ORGET MOM AND APPLE PIE. IT'S an unerring faith in personal transformation that truly defines the American character. We believe foremost in malleability of the people, by the people, and for the people. We hold as self-evident that we can learn to think, feel, and behave differently; that we can change our outlook, remold our character. The pursuit of self-improvement is the national pastime.

That's not to say we think change is easy. On the contrary, we know it is not. And that is exactly why we so admire the qualities required to make it happen: resolve, discipline, mettle. We have learned all too well from our own everyday experiences that sincere attempts at change often breed nothing so much as frustration. Real change is tough.

Real change of what, though? What is it that we're talking about changing when we talk about change? In the broadest sense we seem to mean habitual behavior, and that does cover a lot of ground—from trifling routines so automatic we hardly notice performing them to the mysterious ways of the mind, the ingrained perspectives and beliefs that constitute our personalities. No wonder results may vary.

In fact, sometimes results may not even be definable. Only our conscious struggles to effect change are readily visible. Below the waterline, however, is where more gradual and unconscious change is made—the sum of time and experience—and it is far more difficult to bring into focus.

Habitual behaviors are by definition automatic, and that makes them significant from an evolutionary perspective. Efficiency counts in matters of survival. Making choices, weighing decisions, exercising our power of reasoning ... they all take energy. Repeating behavior reinforces neural pathways, which in turn makes that behavior as effortless and undistracting as possible (it's no coincidence we use metaphors like "groove" and "rut" to describe this process). Habits take a load off our minds, freeing them to do more lofty work. As William James wrote in his 1890 *Principles of Psychology*: "If practice did not make perfect, nor habit economize the expense of nervous and muscular energy," ours would indeed be "a sorry plight."

The downside of this efficiency is that once locked in, habits are hell to break. And that is especially the case with those that rope in the reward systems of the brain. Our neurochemical pleasure pathways make sure that once we find something we like to do, we keep doing it. Transforming these behaviors, then, means fighting the brain's natural orientation. And while that is certainly

possible, it is rarely painless. "Once formed, habits require little attention or intention," says psychologist Roy Baumeister, coauthor of the 2011 book *Willpower,* "but they are costly to change."

As Baumeister and his colleague Mark Muraven once phrased the dilemma: "To do or not to do: which requires more effort?" You'd think, of course, that performing a behavior would necessarily expend more energy than not performing it. The reality, however—as any dieter can attest—is that it's the refraining that's draining. Baumeister and Muraven conducted experiments to better understand the "strength, energy, or other inner resource" required to restrain oneself from surrendering to habit. What they found was evidence that suggested self-control acts not unlike a muscle. "People have only a limited capacity to control and alter their behavior," they wrote, "and this capacity appears to be vulnerable to depletion in the aftermath of strenuous use." Furthermore, they found, if you squander self-control reserves on pointless tasks, you risk a breakdown of resolve in other areas. On the other hand, as with a muscle, working out your self-control over time will build it up.

Self-control is only one personality trait among many, though, and a relatively overt and actionable one at that. Is the rest of the human psyche as amenable to change? Scientific investigation into some of its other components hasn't yet reached any neat consensus, but researchers such as Stanford University's Carol Dweck believe it is only a matter of time. "Personality is a flexible and dynamic thing that changes over the life span and is shaped by experience," she says.

Even our own views about personality—whether we consider specific traits to be fixed or malleable—appear to be subject to change. Consider some of the qualities that we see as character assets: integrity, humility, forgiveness, gratitude, loyalty. At first we might be inclined to see them as natural, innate, and inviolate. But think about it. Don't you really believe that these traits can be cultivated over time (some more easily than others, but still)? In fundamental ways these traits are nothing more or less than personal beliefs, about the world and oneself. And as Dweck has written, "Beliefs matter, beliefs can be changed, and when they are, so too is personality."

We use our beliefs to build personal narratives, the private stories of who we are, where we've been, how we got here, and where we are headed. "Sometimes patterns become ingrained in childhood," says University of Virginia psychologist Timothy Wilson, "and our narratives can become, if not set in stone, at least habitual in some sense." The good news is that they may not be quite as hard to undo as we might assume. As Wilson suggests, "Tweaks to our narratives can sometimes bring about lasting change." Revising one's internal story, in other words, can be like changing a habit. And that, says Wilson, can be done by taking small steps toward acting the way we want to be. Fake it until you make it. "It's a reinforcing cycle," he says. "The better you get at it, the more effective you will be at changing your story."

Understanding one's personal narrative, though, requires a measure of self-knowledge, and too often that is in short supply. Luckily, we are surrounded by potential wellsprings of information about ourselves:

{ Optimist or Pessimist?

If we view our personality traits as habitual patterns of behavior, we can see that they are actually beliefs about the world and our place in it. That can be a good thing or a bad thing.

other people. (Not coincidentally, they can also provide motivation or pressure to change.) "By carefully observing how other people view us, and noticing that their views differ from our own, we could revise our self-narratives accordingly," write Wilson and colleague Elizabeth Dunn in a paper on the limits and value of self-knowledge. And yet it may be hard to find reliable narrators in our social circles too. It isn't surprising to researchers—and shouldn't be to any of us—that, as Wilson and Dunn posit, "people often disagree with their peers about their own personality traits." One person's view of her agreeable nature may be quite different from how agreeable her friends think she is.

Not all change requires such a collective effort. Or any effort, really. Researchers who have studied how personality and values change over a lifetime have isolated a few general trends: (1) people tend to become less open to new experience, but more conscientious, reliable, and agreeable; (2) neuroticism wanes in women, but not in men; (3) extroversion declines in women, but not in men. What drives these trends seems to be the generational changes brought about by milestones and circumstance: marriage and children, work and responsibility. Different times call for different measures.

Oddly, this natural personality drift is almost always less than obvious to the individual living it. In January 2013 Harvard researchers Jordi Quoidbach and Daniel Gilbert, along with Wilson, published the results of a series of online studies that in all included more than 19,000 participants. Adults between the ages of 18 and 68 scored themselves on basic personality traits such as extroversion, emotional stability, and openness to

Passive or Aggressive?

Understanding the stories we weave about ourselves requires self-knowledge. If that's in short supply, we can always consider the opinions of others. But what if they don't see us the way we see ourselves?

new experiences. They were then asked to score themselves again, this time from the perspective of the person they were 10 years ago, and then a third time as the person they imagined they would be 10 years in the future. The three sets of answers plotted how the participants felt they had changed and would change over time, and what they revealed was that no matter the age, people say that while they have changed over the past decade, they do not expect to change in the coming one. A 38-year-old would look back and conclude he had definitely changed from when he was 28. But he would also predict he would be essentially the same at 48 as he was today. A 48-year-old felt the same way: that she had definitely changed from when she was 38 but would change no further in the decade when she would reach 58. Ditto for folks at 58 and 68. Each succeeding age group proved the predictions of the earlier age groups wrong.

Researchers call this the "end-of-history illusion." We apparently believe the present marks the point at which we've finally stopped changing and become the person we will be forevermore. There are a number of reasons we might want to see things this way. For some, it may be comforting and satisfying to believe we are at long last complete. Others might just find it easier to assume they won't change that much than to contemplate a vast, unknown future. Either way, the end-of-history illusion suggests that the power and possibility of self-improvement will remain fraught for us throughout our lives. Which doesn't mean it's not worth the trouble. In fact, it might be the only thing that is. As Kurt Lewin, the father of social psychology, once said, "If you want to truly understand something, try to change it."

Once Upon a Screen

If we let our youngest kids get too attached to those mesmerizing devices, they may have a hard time growing attached to each other later.

By Sherry Turkle

HILDREN WATCH THEIR PARents play with shiny objects all day. While breast-feeding, mothers hold the shiny objects up to their ear. When fathers take their toddlers to the park, those shiny objects hoard so much adult attention that the children, at best, grow jealous and, at worst, go unattended (playground accidents are up).

As soon as children are old enough to know what they want, they angle to get their hands on those shiny objects too. And it is rare that their parents rebuff them. The "passback" has become one of the defining moves of our age—that exasperated handoff from the front seat of the car to the cranky toddler in the rear.

Children of all cultures have always lusted after the objects of grown-up desire. And so shiny objects make their way not only into back seats but also into playpens and cribs and onto playgrounds. Often, our phones and pads, tablets and computers take the place of their blocks and dolls and books. We can understand why children are as drawn to these screens as we are. The screens are mesmerizing. They offer an infinite array of worlds and an immediate connection to other people. Not just fun, they are also legitimate tools of artistic

creation and educational worth. They are compelling.

Screens make children three magical promises that seem like gifts from the fairies: You will always be heard. You can put your attention wherever you want it to be. And you will never have to be alone. From the youngest age there is a social-media account that will welcome you. From the youngest age there is a place where you can be an authority, even an authority who can berate and bully. And there is never, ever a moment when you have to quiet yourself and listen only to your inner voice.

We are embarking on a giant experiment in which our children are the human subjects. There is much that is exciting and thrilling here. But these objects take children away from many things that we know from generations of experience are most nurturant for them. In the first instance, children are taken away from the human face and voice, because people are tempted to let the shiny screens read to children, amuse children, play games with children. And they take children away from each other. They allow them to have experiences (texting, i-chatting, indeed talking to online characters) that offer the illusion of companionship without the demands of friendship, including the responsibilities of friendship. So there is bullying and harassment when you thought you had a friend. Indeed, relationships built in a weightless cyberspace can be badly misconstrued, particularly for children who are often surprised by

hostility where they thought they had alliances. And there is often quick, false intimacy that seems like relationship without risk because you can always disconnect or leave the "chat."

No matter how intriguing or interesting, online connections are not substitutes for the complexities and nuances of face-to-face conversations. Yet one can become so accustomed to what the online world offers—the chance to edit oneself, to present oneself as one wishes—that other kinds of contact feel intimidating. And indeed, many plugged-in children grow up to fear conversation. In my research I often ask children, "What's wrong with conversation?" By about the time they are 10 years old, they can articulate their qualms. To paraphrase: it takes place in real time, and you can't control what you are going to say.

And they do have a point. Of course, particularly for a maturing child, that's also what is so profoundly right with conversation. Children use conversation to practice strategies for dealing with others. Just as important, they use conversation to learn that in real life, practice never leads to perfect. Problem is, perfect is the goal in too many online simulations, and children who have come of age in those worlds may very well be wary in the domain of human relationships where control is not the point.

Imagine an 8-year-old boy in a park, his back against a large tree. He is totally engrossed in his small tablet computer, a recent present. He plays a treasure-hunt game that connects him with a network of other online gamers all over the world. The boy bites his lip in concentration as his fingers move ceaselessly. He doesn't look up. Although he is connected in the game, in the park he is very much alone. Worse, he is unavailable to the invitations of any other child there, to have a catch, maybe, or climb the monkey bars. How will he learn to accept one of those introductions—or make one himself? How will he figure out how to ask questions of other children and listen to their answers?

Actually, he's missing out on even more than that. Children use conversations with one another to learn how to have conversations with themselves. And that capacity for self-reflection is the bedrock of successful child development. The magnetic power of the screen discourages that exploration; the screen jams that inner voice by offering continual interactivity or continual connection. Unlike time spent with a book, when one's mind can

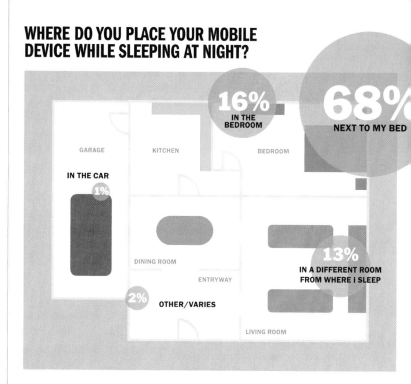

Staying Plugged In

WHERE DO YOU PLACE YOUR MOBILE DEVICE WHILE SLEEPING AT NIGHT?

16% IN THE BEDROOM

68% NEXT TO MY BED

GARAGE
KITCHEN
BEDROOM

IN THE CAR
1%

DINING ROOM

ENTRYWAY

13% IN A DIFFERENT ROOM FROM WHERE I SLEEP

2% OTHER/VARIES

LIVING ROOM

22% of respondents say they screen almost all their calls and tend to reply to voice mail via text or e-mail.

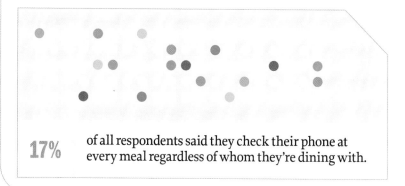

17% of all respondents said they check their phone at every meal regardless of whom they're dining with.

wander and there is no constraint put on reflection, the experience of online apps and games brings children back to the task at hand. That 8-year-old engrossed in his treasure hunt? He has gained mastery over a rule-based game, but there is a loss. He didn't get to hang from his knees on a jungle gym, contemplating the abstract patterns in the upside-down winter sky.

Whereas screen activity tends to rev kids up, the concrete worlds of modeling clay and paints and building blocks slow them down. The particular tangibility

GRAPHICS BY CARL DETORRES

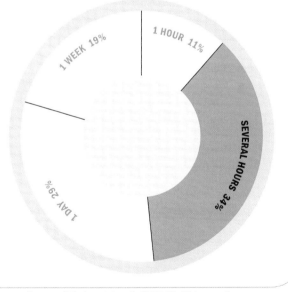

HOW OFTEN DO YOU CHECK YOUR MOBILE DEVICE? (U.S.)

LESS THAN EVERY 5 MINUTES 4%

10 MINUTES 14%

EVERY 30 MINUTES 19%

ONCE AN HOUR 17%

A FEW TIMES A DAY 38%

HOW LONG COULD YOU GO WITHOUT IT? (U.S.)

1 HOUR 11%

1 WEEK 19%

SEVERAL HOURS 34%

1 DAY 29%

Responses of "other" are not shown. The TIME Mobility Poll, in cooperation with Qualcomm, was conducted with 4,700 respondents online and 300 by phone in eight countries, June 29 to July 28, 2012.

of those materials—the sticky thickness of clay, the hard solidity of blocks—offers a very real resistance that presents children with time to think, to use their imaginations, to make up their own worlds. And as the youngest children often work on these kinds of tasks alone, they learn to experience solitude as pleasurable. That capacity for solitude is something that will stand them in good stead for the rest of their lives.

Which is why my greatest misgiving about our lives

An 8-year-old engrossed in an online treasure hunt has mastered a rule-based game. But he didn't get to hang from his knees on a jungle gym, contemplating the upside-down winter sky.

on the screen concerns digital connectivity's third seductive promise: you will never have to be alone. A new generation of adults has grown up afraid to be alone. At red lights and supermarket checkouts, they can't stop themselves from reaching reflexively for a device. Our screen-obsessed lives seem to have left too many of us with a need for constant connection—and a simmering panic at the prospect of ever not being plugged in. Our devices actually encourage a misleading sense of connection, as our never-aloneness is populated with acquaintances who aren't really there at all. And yet we are evolving into a culture that is forgetting every other way to fill downtime besides our screens.

Solitude is where we find ourselves so that we can reach out to have relationships in which we genuinely recognize other people. Those who are unable to be by themselves are liable to behave as if other people were spare parts to support their fragile psyches. Strong friendships of mutual respect blossom from a learned capacity to thrive in a healthy solitude.

In our still-recent infatuation with our mobile devices, we seem to think that if we are connected, we'll never be lonely. But in fact the truth is quite the opposite. If all we do is compulsively connect, we will be more lonely. And if we don't teach our children to be alone, they will only know how to be lonely.

Sherry Turkle is a psychologist, professor at MIT, and author, most recently, of Alone Together: Why We Expect More From Technology and Less From Each Other. *This essay originally appeared on the website Edge.org.*

My Theory of Humor

*Have you heard the one about the guy who thinks you can
learn how to be funny? It goes like this ...*

BY JOEL STEIN

I believe humor can be taught, just like math,
which is strange since I was never able to learn math. But I have sat in
sitcom writing rooms with a dozen other people as we honed our simple
punchline arithmetic: list two similar things, then end the series with the
opposite, the extreme, or the illogical. All comedy is just a refinement on
saying something and then yelling "Not!" This is why funny people are
threatened with violence so often.

I first thought about this in 11th grade, when my friend Ross Novie and I
decided to categorize every type of joke, starting with the "dead-bird genre,"
in which you overshare because you falsely believe everyone has had your
particular weird experience. As in "You know when you're walking down
the street and you see a dead bird and you pick it up and eat it?" Or "You
know when all the other teenagers are hooking up but no one finds you
attractive so you sit at home and try to come up with categories for jokes?"

FROM LEFT: DAVID FISCHER/GETTY IMAGES; STEVE WISBAUER/PHOTODISC/GETTY IMAGES

I think we become funny because we have no other way to attract attention. Models aren't funny.

Athletes aren't funny. Guitar players aren't funny. There has never been a funny rich person, which can be easily verified by a re-viewing of the movie *Arthur*. Funny people constantly work at being funny; in conversations, most of their brain is focused only on listening for the opportunity to interrupt with a joke. Funny people are such bad listeners that the idea of one helping in a hostage negotiation is so preposterous there's never even been an Adam Sandler movie about it.

So deeply do I believe in the teachability of funniness that I once audaciously agreed to lead a class at Princeton on humor writing despite knowing that the class was full of Princeton students. Those kids went on to become Jonah Hill, Lena Dunham, Seth Rogen, Steve Martin, and Mark Twain. OK, not really—but under extreme duress they can now all tell a fart joke.

Still, my inability to create any hilarious writers did shake my faith in the nurture theory of humor. So I decided to get some comedians to share their thoughts, which also happened to be how I killed two of the three hours of that Princeton class each week. "Some people are just born funnier," says Mike Farah, the president of production for Funny or Die. "Then those people find like-minded people, and their natural sense is made even sharper—before it is left to an overwhelmingly unfunny populace to decide just how funny they really are." This is not only his opinion, it is his website's business plan.

Andy Borowitz, who writes the satirical Borowitz Report for *The New Yorker* and won the first National Press Club award for humor, specifically credits birth order. "As the youngest in my family, I wasn't taken seriously; more often than not I was ridiculed. To survive, I had to play the clown. Anecdotally, lots of comedians I know are the youngest in their families." Anecdotally, this isn't remotely true: my younger sister is a divorce lawyer, the least funny career besides bitter divorce lawyer.

Funny people, I was beginning to see, need to believe humor is genetic; they want to feel unique. Then again, their musings are suspect given that not one of them knows a thing about genetics. John Morreall, at least, is a professor who teaches classes on humor at William & Mary and is on the board of *Humor: International Journal of Humor Research*. "Humor is correlated with divergent thinking—seeing unusual connections between things, multiple meanings for words and phrases, and new approaches to solving problems," he says. "In my research, the people with the least developed senses of humor tend to be those who look at everything practically, like electrical engineers." Electrical engineers, apparently, couldn't even come up with a joke about how many electrical engineers it takes to screw in a lightbulb.

Morreall, though, couldn't tell me the inborn-traits-to-learned-skills ratio that leads to divergent thinking. So I called one of the few who have studied this exact question. Willibald Ruch, a psychology professor at the University of Zurich, the funniest university in Zurich, has been on this case since 1980, writing papers with titles like "Assessing the 'Humorous Temperament': Construction of the Facet and Standard Trait Forms of the State-Trait-Cheerfulness-Inventory—STCI" and "The Nature of Humor Appreciation: Toward an Integration of Perception of Stimulus Properties and Affective Experience." After years of arguing that how much you enjoy jokes is not at all heritable, Ruch says now that a study he recently conducted among twins has found that, in fact, recognizing funny is 28% genetic. But how genetic is *being* funny? He actually has a test to determine that, too. What he doesn't have is the funding to underwrite it. "It's not a highly prestigious topic," he says.

Ruch guesses that, because wit is linked to intelligence and creativity, humor is likely significantly more than 28% genetic. "There may be some reproductive advantage to being a good entertainer," he explains. The unfunny, in other words, will one day be extinct. Which is going to be a big problem when we have trouble with our electricity. That, by the way, is called "a callback." Which someone taught me.

About the Authors

DAVID BJERKLIE is a science writer and the author of children's books on butterflies, agriculture, and environmental justice. Formerly, he was the senior science reporter at TIME, senior editor at TIME for Kids, and science writer/editor at TheVisualMD.com.

LAURA BLUE is a senior contributing health writer for TIME.com. Her particular area of interest is the effect of day-to-day behaviors such as diet, exercise, smoking, and stress on our general health and longevity.

MICHAEL Q. BULLERDICK is a writer and editor whose work has appeared in numerous national magazines and books.

JOHN CLOUD is a journalist who has written for TIME, *The Washington Post,* and *The Wall Street Journal.* He has won a number of awards for his reporting on social issues and health.

HENRY KELLERMAN is a psychologist and psychoanalyst in New York City and the author and editor of more than 20 books. He has held staff positions at New York University, the New School, City University of New York, and the Postgraduate Center for Mental Health.

JEFFREY KLUGER is a senior editor at TIME, overseeing its science and technology reporting. He has written or co-written more than 40 cover stories for the magazine and regularly contributes articles and commentary on science, behavior, and health.

MICHAEL D. LEMONICK is a senior science writer at the research organization Climate Central. He has also covered the subject for

TIME for more than 20 years and is the author of several books on astrophysics, including *Echo of the Big Bang; Other Worlds; The Search for Life in the Universe;* and *Mirror Earth: The Search for Our Planet's Twin.*

BELINDA LUSCOMBE, an editor-at-large for TIME, writes about the science of relationships—at home, at work, or in cyberspace. She also covers the pop-culture world: movies, books, TV shows, and gossip. Luscombe has worked at TIME since 1995, after moving to New York City from Sydney.

ALICE PARK is a staff writer at TIME. Since 1993 she has reported on the frontiers of health and medicine in articles on such topics as cancer, anxiety, and Alzheimer's disease.

JOEL STEIN is a columnist for TIME. His first book, *Man Made: A Stupid Quest for Masculinity,* is in bookstores now.

MAIA SZALAVITZ is a neuroscience journalist for TIME.com and coauthor of *Born for Love: Why Empathy Is Essential—and Endangered.*

SHERRY TURKLE is the Abby Rockefeller Mauzé Professor of the Social Studies of Science and Technology at MIT, and founder and director of the MIT Initiative on Technology and Self. Her latest book is *Alone Together: Why We Expect More From Technology and Less From Each Other.*

BRYAN WALSH is a senior editor for TIME International. He also writes the "Going Green" column for TIME and TIME.com and contributes to the website's environmental-issues blog, Ecocentric.